D0992223

WHAT IS DISEASE?

BIOMEDICAL ETHICS REVIEWS

Edited by

James M. Humber and Robert F. Almeder

BIOMEDICAL ETHICS REVIEWS

WHAT IS DISEASE?

Edited by

James M. Humber

and

Robert F. Almeder

Georgia State University, Atlanta, Georgia

Humana Press • Totowa, New Jersey

The Library of Congress has cataloged this serial title as follows:

Biomedical ethics reviews—1983— Totowa, NJ: Humana Press, c1982—
v.; 25 cm—(Contemporary issues in biomedicine, ethics, and society)
Annual.
Editors: James M. Humber and Robert F. Almeder.
ISSN 0742-1796 = Biomedical ethics reviews.
1. Medical ethics—Periodicals. I. Humber, James M. II. Almeder, Robert F.
III. Series.
[DNLM: Ethics, Medical—periodicals. W1 B615 (P)]
R724.B493   174'.2'05—dc19   84-640015
                        AACR2 MARC-S

Contents

Preface

This book is the fourteenth volume in a series designed to review and update the literature on issues of central importance in bioethics today. Each annual volume of *Biomedical Ethics Reviews* is organized around a central theme. This year the topic for discussion is the concept of disease.

What Is Disease? opens with a lengthy essay by Christopher Boorse. Some 20 years ago, Boorse wrote a series of articles in which he offered what many consider to be the canonical defense of the view that disease is a value-free scientific concept. This view, which has come to be called naturalism or neutralism, gave rise to numerous objections. In his essay,"A Rebuttal on Health," Boorse responds to all those who have criticized his theory, and in the process both clarifies and modifies his early views. Boorse's article is followed by a number of essays in which nonnaturalistic theories of disease are analyzed, developed, and oftentimes defended.

We greatly hope our readers will find the present volume of *Biomedical Ethics Reviews* to be both enjoyable and informative, and that they will look forward with anticipation to the publication of next year's volume, *Issues in Alternative Medicine*.

James M. Humber
Robert F. Almeder

Contributors

George Agich • School of Medicine, Southern Illinois University, Springfield, Illinois

John D. Banja • Center for Rehabilitation Medicine, Emory University, Atlanta, Georgia

Christopher Boorse • Department of Philosophy, University of Delaware, Newark, Delaware

K. Danner Clouser • Hershey Medical Center, Pennsylvania State University, Hershey, Pennsylvania

Charles M. Culver • Florida College of Physician Assistants, Miami, Florida

Bernard Gert • Department of Philosophy, Dartmouth College, Hanover, New Hampshire

Frederik Kaufman • Department of Philosophy and Religion, Ithaca College, Ithaca, New York

Michael Ruse • Departments of Philosophy and Zoology, University of Guelph, Guelph, Ontario, Canada

Stan van Hooft • School of Social Inquiry, Deakin University, Toorak, Malvern, Australia

Mark Woodhouse • Department of Philosophy, Georgia State University, Atlanta, Georgia

Introduction

In this essay Boorse answers two decades of attacks on his "biostatistical theory" (BST), which defines health as the absence of disease and disease as statistically species-subnormal biological part-function. An unrepentant naturalist, Boorse reaffirms all aspects of his analysis, including a later revision making illness as value-free as disease. However, he admits for the first time that the BST may need the concept of a statistically normal environment to handle environmental disease.

Boorse defends his theory against multifarious objections and counterexamples. He replies to charges that it is covertly value-laden, either in specific vocabulary or in its choice of physiological goals. He denies that it presupposes biological essentialism, describing it instead as a populationist view. He re-emphasizes the contrast between theoretical and practical, or pathological and clinical, levels of description, arguing that disease is basically a concept of the first type. But on this scientific foundation, he suggests, value-laden clinical and social concepts of disease—"disease-plus" concepts—can be constructed by adding evaluative criteria. Such a "multi-level" framework, he maintains, can accommodate the richness of clinical and social health concepts while allowing various philosophies of clinical practice.

Finally, Boorse addresses a potpourri of real cases raised by critics, including baldness, hanging, skin parasites, masturbation, mountain sickness, sickle trait, premenstrual syndrome, inflammation, oncogenes, osteoporosis, and senility, as well as many hypothetical cases. He argues that almost all these cases,

1

instead of disconfirming the BST, confirm it. However, he stresses that the medical disease concept has limits of precision, and he concludes that some phenomena (e.g., heterosis) may be best handled by other concepts, such as adaptation.

A Rebuttal on Health

Christopher Boorse

Introduction

Twenty years ago, in four papers,[1] I offered a unified descriptive analysis of health, disease, and function. In recent philosophy of medicine, these papers are often treated as a standard defense of one pole on the spectrum of views about health: the extreme view that, at least at the theoretical foundation of modern Western medicine, health and disease are value-free scientific concepts. Theoretical health, I argued, is the absence of disease; disease is only statistically species-subnormal biological part-function; therefore, the classification of human states as healthy or diseased is an objective matter, to be read off the biological facts of nature without need of value judgments. Let us refer to this general position as "naturalism"—the opposite of normativism, the view that health judgments are or include value judgments. Following Nordenfelt (1987), let me call my specific naturalist theory the "biostatistical theory (BST)," a name emphasizing that the analysis rests on the concepts of biological function and statistical normality.[2]

Regrettably, the BST's influence is hardly due to its multitude of converts. Rather, many writers on health use the BST as a stalking-horse, exposing the defects of my extreme view en route to their own more sensible positions. There are some welcome exceptions. Norman Daniels (1981, 1985), while raising some problems for the BST, gives it an important role in his theory of just health care. Germund Hesslow (1993, p. 8) kindly describes it as "about as close as it is possible to get to a correct explication of what doctors mean by 'disease,'" although his actual thesis is that health and disease are "superfluous" and "irrelevan[t]" concepts in medicine. At the opposite extreme is Engelhardt's opinion that if I succeeded in analyzing anything, it was "not clinical medicine's concept of disease," but one "of primary interest only to biologists" and quite possibly not even to them (Engelhardt, 1986, p. 171). (But at least I "unwittingly" performed the "important service" of underscoring the difference between pure science and clinical medicine.) Of course, I am not the only naturalist. Besides earlier writers, the naturalist camp currently includes Scadding, Kendell, Klein, Szasz, Ross, and to some extent Kass. But normativism is still the received view in philosophy of medicine as it was two decades ago, though the depth of normativist theories has improved in the interval. It is also gratifying to see two normativists (Wakefield, R. Brown) analyzing disease as harmful dysfunction. Such a view, simply by including the key dysfunction requirement, is nearly as good as the BST for clarifying most actual health controversies.

In this article I reply to twenty years of criticism, aiming to meet every important published objection to my views on health.[3] Space does not permit rebuttal specifically on functions, nor much criticism of rival theories of health; these are topics for other essays. As will be seen, I am an unrepentant naturalist: I make here only one new act of contrition, on environmental disease. On the whole, critics' objections seem to fall into the following categories. A few writers, especially those who did not read all four original papers, misinterpret the BST. Some have trouble thinking within a naturalist framework, and so their arguments for normativism turn out to presuppose it. Some critics project implausible views for me on subjects that I have said little about—nosology, disease explanations, medical ethics, clinical style. Some objections were answered in the original papers. A few points can be incorporated in the BST as friendly improvements. Of critics' counterexamples, almost all seem to me better handled by the BST than by rival theories, and therefore to confirm rather than disconfirm it. Whether the example is masturbation, sickle trait, premenstrual syndrome, or unwanted pregnancy, it is the BST, not rival theories, that best explains medical disease judgments and our reactions to them.

Interestingly, many objections seem at bottom to be attacks on the concept of disease, not on my analysis of it. The serious philosophical issues between the BST and its critics are not, I think, about the correct analysis of "disease." Rather, they are about the prospects for a genuine concept of health—individual, nontypological, positive, or of some other kind—that could differ from the absence of disease, and about what medical theory, practice, or social institutions might be based thereupon.

Summary of the BST

Strictly for convenience, I begin with a brief outline of my theory. Since I have written a full introduction elsewhere (CH), I will stress overlooked elements and changes of view, of which I have had one major and several minor ones.

The BST aimed at a clear analysis of that theoretical negative idea of health in scientific medicine which is expressed by its cliché "Health is the absence of disease." Such a thesis requires a very broad usage of "disease," one including injuries, poisonings, environmental traumas, growth disorders, functional impairments, and so on through the vast range of conditions that medicine views as inconsistent with perfect health. Even in medicine, such a broad usage of "disease" occurs mainly in two places: the headings of reference works and general discussions of health. But if narrower meanings are more common, the broad meaning is still legitimate—exactly as legitimate as the principle that health is the absence of disease (HTC, p. 551), since there is no possibility of health's being the absence of disease unless all the above causes of death are included. By CH, however, I realized that this broad usage of "disease," which so upsets many readers, can be avoided if one switches from "health vs disease" to "normal vs pathological." The goal of the BST, then, is to analyze the normal–pathological distinction, which I claim is the basic theoretical concept of Western medicine (CH, p. 365).

After noting that diseases are not always disvaluable nor disvaluable conditions diseases, I argued that other common ideas in health definitions also fail as simple necessary or sufficient conditions of disease: treatment by physicians, statistical normality, pain and suffering, disability, adaptation, and homeostasis. To capture the modern extension of "disease," what seemed requisite was a modern explication of the ancient idea that the normal is the natural—that health is conformity to a "species design." In modern terms, species design is the internal functional organization typical of species members, which (as regards somatic medicine) forms the subject matter of physiology: the interlocking hierarchy of functional processes, at every level from organelle to cell to tissue to organ to gross behavior, by which organisms of a given species maintain and renew their life. The common feature of all conditions called pathological by ordinary medicine seems to be disrupted part-function at some level of this hierarchy. After discussing various details, I offered the following final definition:[4]

1. The *reference class* is a natural class of organisms of uniform functional design; specifically, an age group of a sex of a species.
2. A *normal function* of a part or process within members of the reference class is a statistically typical contribution by it to their individual survival and reproduction.
3. A *disease* is a type of internal state which is either an impairment of normal functional ability, i.e. a reduction of one

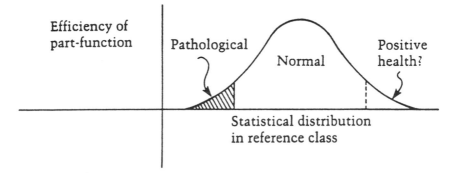

Fig. 1. Normal vs pathological.

or more functional abilities below typical efficiency, or a limitation on functional ability caused by environmental agents.

4. *Health* is the absence of disease.

   In this definition, the reference class was restricted by sex and age because of differences in normal physiology between males and females, young and old. I said that one might also subclassify by race, and Reznek argues for so doing based on such examples as Masai sensitivity vs pygmy insensitivity to growth hormone.[5] The clause on environmental injuries was meant to help handle the second of two classes of recognized diseases that the theory seemed not to cover (structural diseases and universal diseases); on this see Defense Mechanisms in this chapter. The term "normal functional ability" had been defined dispositionally, as the readiness of an internal part to perform all its normal functions on typical occasions with at least typical efficiency. "Typical efficiency" of a part-function, in turn, is efficiency above some arbitrarily chosen minimum in its species distribution. So the basic idea of the BST is shown by Fig. 1 (CH, p. 370).

   This definition of the normal–pathological distinction depends on three perhaps debatable concepts: the reference class, statistical normality, and biological function. For biological function I offered a value-free analysis in a separate paper (WF). A function was a causal contribution to a goal, and the actual goals of organisms were defined in the manner of Sommerhoff (1950). An organism or its part is directed to goal G when disposed, throughout a range of environmental variation, to modify its behavior in the way required for G. (I proposed a slight change to accommodate goal-directedness to currently impossible goals, as when a cat stalks a nonexistent mouse.) On this analysis of goal-directedness, most behavior of organisms seems to contribute to many goals at once: "individual survival, individual reproductive competence, survival of the species, survival of the genes, ecological equilibrium, and so forth" (HTC, p. 556). I suggested that different subfields of biology may use different goals as the focus of their function statements. But since physiology was the subfield on which somatic medicine relies, medical functional normality was presumably relative to the goals physiologists seem to assume, viz, individual survival and reproduction.

The role of Sommerhoff's view of life in inspiring my naturalism about health is overlooked by many critics. Sommerhoff saw goal-directedness as the key feature dividing living organisms from dead or inorganic matter. In a trenchant passage in *Analytical Biology,* he wrote:

> [Except for borderline cases of life,] it would be hard to find any level of organic activity which does not invite us to think of vital activities as being somehow purposive, as being subject to tendencies which are directed towards the fulfillment of specific and mutually interrelated ends. On the phenomenal level from which all science must proceed, life is nothing if not just this manifestation of apparent purposiveness and organic order in material systems. In the last analysis, *the beast is not distinguishable from its dung* save by the end-serving and integrating activities which unite it into an ordered, self-regulating, and single whole, and impart to the individual whole that unique independence from the vicissitudes of the environment and that unique power to hold its own by making internal adjustments, which all living organisms possess in some degree. (Sommerhoff, 1950, p. 6; italics added)

Now a basic difference between normativists and naturalists is that the former view health as a welfare-like concept, while the latter view it as a life-like one.

> [Normativist] writers seem to start from the intuition that diseases are *bad* conditions of the organism—physiological evils, or psychological evils in the case of mental health. On this normativist view, health and disease belong to the welfare-harm family of ethical concepts. The normativist writer's job is to distinguish medical from nonmedical evils. ... Other writers ... seem to begin instead with an intuition that places health and disease in the life-death family of biological concepts. (CH, pg. 366)

I suspect that one cause of this difference is normativists' failure to ponder Sommerhoff's beast–dung point. Many normativists seem never to have asked themselves what distinguishes living from nonliving things, and so biology, the science of life, from chemistry and physics. In ignoring this question, they discount the possibility that the life–nonlife contrast may automatically generate a health–nonhealth contrast as a concept of the same scientific family.

Still, it is worth noting that my analyses of health and function are separable, in that one could ground the BST on a different analysis of function. For example, one could join my health–disease definition to some version of Wright's "etiological" approach (1973), the most popular idea in the function literature. Wright suggested that the functions of a thing are those of its effects which historically explain its presence. In particular, the functions of a part of an organism are the effects which, through evolution, fixed it in the population. Since evolutionary advantage comes from an individual's inclusive fitness, and I named physiologists' goals of interest as individual survival and reproduction, our two analyses almost always agree on biological part-functions. But they can diverge, for example, where a trait was established by an effect no longer serving the organism's goals in current environments. Several BST critics cite this phenomenon. However, it is possible that their objection is to my analysis of function, not to my analysis of health, and would disappear on Wright's function theory.[6] Health concepts based on the two analyses of function will also differ on group-related functions, if any exist, fixed by group selection. In any case, I may be right about health and wrong about functions; or wrong about health and right about functions; or simultaneously right or wrong about both.

The above account defines a theoretical concept of health, not a practical one. It aims at a pathologist's concept of disease, not a clinician's, and still less at any social or legal category. Originally, then, I invoked a contrast between disease and illness to distinguish value-free theoretical health from a value-laden practical counterpart, freedom from illness. I said that illness was only a subclass of disease, including diseases serious enough to have certain normative features.

> A disease is an illness only if it is serious enough to be incapacitating, and therefore is
>   (i) undesirable for its bearer;
>   (ii) a title to special treatment; and
>   (iii) a valid excuse for normally criticizable behavior
> (DI, p. 61)

I then applied this distinction between disease and illness to mental health.

Unfortunately, though the idea of value-free theoretical vs value-laden practical concepts of health still seems to me sound, I now believe that I chose the wrong concept (illness) to illustrate this distinction. I changed my mind for the following reasons. First, one main reason for considering "illness" as value-laden was that it cannot apply to animals and plants. However, there seems to be no semantic difference between "ill" and "sick," yet we do often call animals, and even plants, sick. The two terms seem to differ only in that "ill" is more elevated in tone, possibly because it is more common in England, where things have greater dignity. But in that case, the reason we do not call sick animals "ill" is the same reason that we don't call dead animals "deceased."

I drew the wrong conclusion from my fruit flies (DI, p. 56): they were sick and are now dead, but it would merely be comical to say they were ill and are now deceased. Secondly, and more importantly, the difference between disease and illness does not, in fact, seem to be a difference of severity. Usually when I am sick, I am less disabled from my normal activities than I would be if blind or paraplegic. Yet blindness and paraplegia are not illnesses, while a mild case of flu is. Why? Not because illness must be a process, since the same point could be made with severe progressive blindness. Rather, it is because "sick" (or "ill") refers to systemic rather than local disease, to disease which in some sense incapacitates by permeating the whole organism, as do infectious diseases via blood-borne substances and disruption of central homeostasis. But this term "systemic," though vague, is a physiological or pathological term, not an evaluative one. Consequently, I now consider "disease" and "illness" equally value-free. At worst, "illness" is value-laden if the degree of (systemic) incapacitation required for illness involves evaluative choice; but some have urged this thesis regarding "disease" as well.

Accordingly, in CH I replaced my too-simple disease–illness distinction with a range of "grades of health" (Fig. 2). In illness's former role of value-laden practical concept, I now placed two new ones. Diagnostic abnormality was a clinically apparent pathological state. Therapeutic abnormality was a diagnostic abnormality meriting treatment. Diagnostic normality is already value-laden because, among other reasons, what diagnostic tests are justified depends on risk, cost, and benefit. Obviously therapeutic normality is an ineluctably normative concept. Any treatment of persons whatever is open to moral review (cf. DI, p. 54), including giving diagnostic tests, and can only be justified by value judgments. Moreover, there are specific values traditionally deemed essential to medical practice—the ones in medical-ethics texts—and good medical practice also requires a kind of interpersonal sensitivity to which BST critics sometimes refer (*see* Clinical Implications). Contrary to some critics, I have never

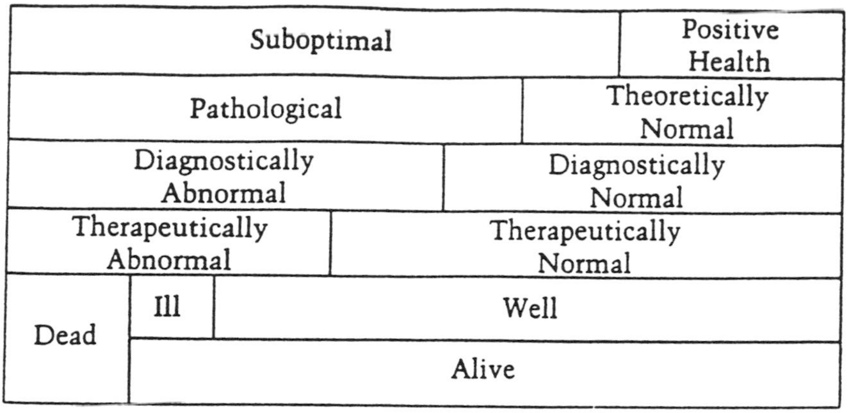

Suboptimal				Positive Health
Pathological				Theoretically Normal
Diagnostically Abnormal			Diagnostically Normal	
Therapeutically Abnormal			Therapeutically Normal	
Dead	Ill		Well	
	Alive			

Fig. 2. Grades of health.

doubted that medical practice is permeated by values, nor that a good doctor must have more tools than a scientific knowledge of pathophysiology. I have not spoken to issues of clinical sensitivity for lack of anything original to say. But I did insist on the value-ladenness of medical practice—originally by the distinction between disease and illness, now by the distinction between disease and a richer family of practical concepts that could easily be extended beyond the two in Fig. 2. The diagram is limited, for example, to concepts of clinical medicine. It does not include socially defined categories, such as employment disability or legal insanity. Finally, CH distinguished two kinds of medical practice, which I called "core" and "peripheral" medicine but might equally have called "therapeutic" and "nontherapeutic." To me nontherapeutic medicine (contraception, circumcision, cosmetic surgery, abortion, euthanasia) is not necessarily illegitimate, but it is more controversial than the core practice of health promotion.

Virtually the whole BST is applicable to mental health, having in fact been developed as a tool for clarifying psychiatric issues. But since most critics focus on somatic medicine, I will only briefly list psychiatric implications. First, theoretical mental health exists, quite literally, on one condition: "that human psy-

chology be divisible into part-processes with biological functions" (CH, p. 376). Here "biological" does not mean "physico-
chemical," an unfortunate confusion persisting in some BST
criticism. Biology, by definition the science of life, is in fact "the
study of inherited functional structures of organisms produced by
evolution," and as such "embraces … species-typical psychology
as much as species-typical physiology" (CH, p. 376). I cited classical psychoanalysis as an example of what kind of personality
theory would, by anatomizing the mind into part-functions, support a genuine theory of mental health. Independently of psychoanalysis, in the last two decades many psychologists have regained
interest in "evolutionary psychology," a major topic in Darwin.[7]
Not only is this trend scientifically admirable, but it also offers,
as Wakefield's three papers show, just the style of thinking
required for a genuine theory of mental health.

I argued in MH, as have others, that American psychiatry
has often abandoned the effort to base its nosology on a substantive theory of normal mental part-functions, instead using personal or social values to classify people as sick. Although, in the
wake of the homosexuality controversy, DSM-III and its succes-
sors did abjure diagnosis by social values, the new nosology also
used an atheoretic approach attacked by many as confused and
sterile.[8] Nevertheless, it seems inevitable that at least major mental
disorders (the psychoses) involve genuine biological part-dys-
functions of the mind. Thus, while some feel that "disease" has
inescapable physical connotations and so prefer not to speak of
"mental disease," the term "psychopathology" looks literally justified for at least some conditions in psychiatry's domain. Thus,
one can reject the main thesis of "anti-psychiatrists' like Szasz,
while still endorsing many of their charges against past psychiatry. Finally, I suggested several obstacles to applying practical
(as opposed to theoretical) health concepts to mental disorders
(DI, pp. 62–68), concluding that none of my three stated conditions on "illness" seemed "to apply automatically to serious mental
diseases" (DI, p. 66). Some of this analysis is now obsolete in

light of my change of view on "illness." But the points about moral or legal responsibility of the mentally ill remain important, as does the conclusion that one cannot expect mental illness to be "just like any other illness" even as to therapy, let alone for social institutions like criminal law. And a new problem now arises about "mental illness": the issue whether the "system" in systemic incapacitation should be the whole organism or the whole mind (cf. Champlin, 1981, p. 478).

Before I answer criticisms of the BST, here is a joint list of its strengths as seen by me, Reznek (1987, pp. 127, 128), and Nordenfelt (1987, 21–23).

1. It explains the divergence between facts about disease and facts about desirability or treatability (HTC, Reznek).
2. It offers a unified treatment of extremal diseases (HTC, Reznek).
3. It explains the phenomenon of symptomless disease (Reznek), and indeed how health judgments can be independent of gross output (HTC).
4. It explains disease judgments about plants and animals (HTC, Nordenfelt, Reznek) and why what looks like disease in an organism may not be so if there is no malfunction (Reznek).
5. It applies to both physical and mental health (Nordenfelt, MH, CH).
6. It gives a clear account of logical relations among important health concepts (HTC, Nordenfelt).
7. It gives health concepts scientific status (HTC, Nordenfelt), and in the process simultaneously explains cultural variation in disease judgments while avoiding cultural relativism (HTC, Reznek).
8. It explains the partial successes of other analyses (HTC).

"In spite of all these virtues," writes Reznek, "this account of disease and of pathological conditions is mistaken" (Reznek, 1987, p. 128). How can this be?

Rather than taking critics one at a time, I have tried to organize their criticisms in conceptual categories. But it was sometimes convenient to pursue one critic more systematically, so I hope readers will forgive me if the discussion occasionally gets a bit ahead of itself.

Technical Objections to the BST
Circularity

At least two authors charge me with circularity. Scadding (1988), mostly a strong ally as a fellow naturalist, notes two points of disagreement. One runs as follows:

> [Boorse's] identification of disease with an "internal state" is confusing. This state is presumably regarded as the explanation of the phenomena constituting a disease; but at the same time, it is itself implied from these phenomena. (Scadding, 1988, p. 123)

This complaint is made against the background of Scadding's own definition of disease as

> the sum of the abnormal phenomena displayed by a group of living organisms in association with a specified common characteristic or set of characteristics by which they differ from the norm of their species in such a way as to place them at a biological disadvantage (Scadding, 1988, p. 121)

and also his point that

> The simple idea that diseases are causes of illness and that diagnosis consists in identifying the disease which is causing the patient's illness is obviously erroneous. ... (Scadding, 1988, p. 118).

That idea is erroneous because the diagnostic name ("chronic bronchitis," "asthma") may be only a summary of a group of symptoms, functional abnormality, and so on, and therefore fail to explain what it describes. But I see no inconsistency between that point and my definition of disease. I do not claim that all diagnoses are explanatory rather than descriptive; I have said almost nothing about the form of disease explanations. And disease as an "internal state" is consistent with Scadding's "sum of abnormal phenomena" if none of these phenomena is external to the organism. The real contrast between my analysis and those of Scadding and Kendell is over Scadding's second issue, whether part-dysfunction is necessary and sufficient for disease *(see* Dysfunction vs Diversity of Disease Classification).

Bechtel begins another charge of circularity as follows:

> Ultimately, Boorse appeals to a standard of normality to distinguish healthy from diseased functioning. First, however, he gives the impression of trying to explain what normal is. He cites a statement from King that appears to define normal in terms of the design of the organism. ... (Bechtel, 1985, pp. 142, 143)

The King quotation was: "The normal ... is objectively, and properly, to be defined as that which functions in accordance with its design." Bechtel then says that "because the organism is changing under selection pressures," its design cannot be an internal essence. My "way out" of this dilemma is my goal-based empirical theory of statistically normal functions. However,

> Boorse has rather clearly moved in a circle now, for normal is being used to define functioning according to design, which was to explicate the notion of normal. (Bechtel, 1985, p. 143)

If I understand this criticism, it may rest on some confusion of medical and statistical normality. I called C. Daly King's statement an "admirable explanation of clinical normality." I should have said "medical," not "clinical," since my basic analysis really aims at the pathologist's, not the clinician's, concept of disease. But I see no circle. What I proposed was to explicate medical normality as statistical normality of function (or more accurately, statistical nonsubnormality of function). Medical normality, as King said, is functioning according to design. But the species design is, in fact, simply those functions statistically typical in species members. Given a focus on functions, medical normality and statistical (nonsub)normality are the same thing. But one must add the function concept to the statistical-normality concept to get medical normality, so the two kinds of normality differ. If any circle remains here, then perhaps the King quote is not as admirable as I thought. In any case, HTC, my careful version of the basic analysis, neither cites Daly King nor uses the term "species design" in any definition.[9]

Bechtel (like HTC) goes on to note that the above idea makes universal diseases impossible, and that I proposed an environmental clause to handle this problem. He quotes my disjunctive summary from DI and complains:

> (If this modification is to solve the above problem [of universal diseases "such as caries"], Boorse needs to amend this to hold that a disease is an abnormality that is atypical *and* caused by the action of a hostile environment.) To make this definition work, however, we need to know what actions are hostile, something we can only judge once we know the state they are disruptive of and hence hostile to. (Bechtel, 1985, p. 143)

The logical criticism here mystifies me. A universal disease could not be "atypical," so Bechtel's conjunctive version would not help at all with universal disease, which was the only reason for introducing an environmental clause. As for a "hostile" environment, it is one that disrupts something in the hierarchy of statistically species-normal part-functions, i.e., reduces a part's typical contribution to a physiological goal (*see* Specific Terms)—that is the "state," as explained in WF and HTC.

Vagueness

Many critics can be seen as complaining about vagueness in key concepts of the BST. Hare (1986), especially, raises a long series of puzzles about disease, first just for fun and then to attack the BST. He concludes with a master argument for normativism based on the example of head hair vs leg hair. Except for the hair examples, Hare's criticisms of the BST are to some degree cast as charges of vagueness in key concepts: "internal," "state," "species," "typical," "environmental cause," and "characteristic of the organism's age." Hare's examples are considered in a later section under separate headings, along with other authors raising similar themes.

Strictly speaking, there is no such charge as vagueness, since vagueness is ineliminable. All ordinary concepts, perhaps all concepts, are vague in having borderline cases. What one can charge is that a definition is too vague by some standard or for some purpose, e.g., clinical description. The standard I accept is medicine's, and in particular pathology's, actual considered usage of the term "pathological." My line on vagueness will be that the BST is never less precise than medical usage and fits all its clear cases, except the two classes mentioned in HTC. Occasionally the BST may be more precise than medicine, in which case it is a precising definition. Usually, the BST seems to me simply to match the vagueness of medical usage, leaving indeterminate conditions of which I believe the disease status is indeterminate in medicine. But it can be an important virtue of a definition to explain the limits of a concept's resolving power. Indeed, this has been one of my main themes, that concepts of health cannot do everything people want them to do. At any rate, whether critics mean to suggest that my analysis is too vague to classify medically clear cases or that it clearly misclassifies them, I shall argue that they are mistaken. The real threat to the BST is, of course, cases where it and medical usage clearly diverge.[10]

Covert Normativism

From two different directions, several critics allege that the BST is not, in fact, a value-free analysis.

Specific Terms

Chapter 3 of Fulford's *Moral Theory and Medical Practice* (1989) is mostly a long debate. It is billed as a "debate about Boorse's version of the medical model ... between two imaginary philosophers, one broadly descriptivist in outlook, the other broadly non-descriptivist" (p. 37). However, this billing is misleading, since what actually happens is that both philosophers immediately reject my position and go on for 19 pages from there.

The rejection occurs when "nondescriptivist" and "descriptivist" agree, from their first interchange, that I constantly slip from descriptive to normative terms in characterizing disease. Their first example (from DI, p. 59) is that I define disease first as a "deviation" from "the natural functional organization of the species" and only two lines later shift to "deficiency." While "deviation" could be descriptive, "deficiency" is normative, Fulford's debaters agree. The descriptivist, though apparently meant to be my champion here, states the general charge:

> Boorse's *claim* notwithstanding, he continues to use both 'dysfunction' and 'disease', and the ideas expressed by these terms, evaluatively. (p. 37)

The debaters quickly find further examples: "hostile" environment (DI, p. 59), "interference" with a mental function (MH, p. 62), biologically "incompetent" behavior (MH, p. 76), "disrupted" cognitive functions (MH, p. 77), and neuroses "blamed" on an "injurious" cultural environment which fills children's minds with "excessive" anxieties, "grotesque" role models, and "absurd" prejudices (DI, p. 65).

Admittedly, my rhetorical seas rose a bit high in that last passage, but Fulford's charge seems to me unfounded. Some terms that he calls normative are not or need not be so. In general, Fulford has a rather expansive view of the normative, counting, e.g., "kidney" as a value-laden term.[11] But where my terms are normative, they can easily be replaced by nonnormative ones; the evaluative rhetoric is eliminable without loss. Fulford misses this because—having apparently looked only at DI and MH, not WF or HTC—he pays no attention to my empirical view of functions based on Sommerhoff goal-directedness. From that view, in every case it is, I think, easy to see how to restate the point in descriptive language.

Thus, even if we grant *arguendo* that "deficiency" is norma-tive, the function theory imposes a natural polarity on physiologi-cal processes.[12] The function of a physiological process is its contribution to physiological goals. By "deficiency" of function, then, I mean simply less function, less contribution to the goals, than average. This is an arithmetic, not an evaluative, concept. An easy example is homeostatic functions like regulation of body temperature or of blood pressure, gases, or electrolytes. Insofar as what regulates these variables is directed to the physiologic goal of constancy, deficient function is simply much greater varia-tion, much wider swings, than average, or failure to return from one at all. In general, whenever one knows the goal of a process, one knows what is more or less function, and "deficiency," in the context quoted, simply means much less than average. I am not sure that "hostile" is evaluative at all. Surely it and its synonyms ("aggressive") are common enough in biology.[13] A hostile per-son, like an enemy, is someone who is trying to harm you. In physiology, harm is lowered functional efficiency in serving physiological goals. "Interference" is used even in physics, and many sciences talk of "disrupting" an equilibrium. To interfere with or disrupt a functional process is, for me, to lower its degree of achievement of physiological goals empirically given by the Sommerhoff analysis. If the goals aren't achieved at all, the functioning is incompetent. So it seems to me that the terms Fulford complains of, insofar as they are evaluative, are easily eliminable.[14]

Since Fulford's long, philosophically sophisticated debate, like the rest of his book, is very much worth reading, I must note two other ways in which it misleads about my own views. Fulford uses "descriptivism" in the usual philosophical sense, for the general view about value terms holding that they can be descrip-tively defined. But then descriptivism/nondescriptivism is a poor framework for debating the BST. The basic issue for the BST is normativism about health: whether health and disease, specifi-cally, can be descriptively defined. The position that they can I am calling naturalism. Naturalism about health could only be a kind of descriptivism, on Fulford's usage, if health is a value concept, which is exactly what I deny. I have always taken "normative" and "descriptive" for contraries, believing that no value term can be descriptively defined. In Fulford's terms, I have never drawn a descriptivist breath. But the structure of his debate creates the impression[15] that the descriptivist is my defender, whereas, as it proceeds, I often agree with his nondescriptivist.

Apparently, Fulford cast the debate as descriptivism vs nondescriptivism, rather than normativism vs naturalism, because he views normativism about health as nearly axiomatic. At bottom, he cannot grasp how anyone could fail to be a normativist, and he thinks that I am one in spite of myself. Thus, during the debate, his debaters seek to construct the most plausible version of the BST they can imagine, which must be one conceding the initial normativist thesis of his chapter that health and disease are value-laden ideas. But why is Fulford so convinced of normativism? In part, surely, because he ignores the Sommerhoff-based goal-theory of functions on which the BST actually rests. But he seems also to be misled by etymology. The chapter begins:

> The question ... whether the terms 'disease' and 'dysfunction' are really value-free as used in technical contexts in medicine ... is an important question. ... That it should be raised at all, however, is due to a property which 'disease' and 'dysfunction' have in common with 'illness'—namely, that although etymologically value terms, and although capable of being used as such, they may also be used (as in technical contexts in medicine) with mainly factual rather than evaluative connotations. (p. 35)

This framework excludes my view from the start, which is that theoretical health and disease in medicine are wholly, not "mainly," factual. But as to etymology, quite apart from its general weakness in semantics, Fulford offers no proof that "disease" is etymologically evaluative in the first place. There are hundreds of English words in which the prefix "dis" has no evaluative meaning, but instead conveys undoing or separation: "disconnect," "discover," "disgorge," "disrobe." A disrobed woman is an unrobed woman, not a poorly robed one. Perhaps Fulford sees the etymological value in the "dys" of "dysfunction" but the "ease" of "disease." But one should not be too hasty even here. Although what makes a person feel easy or uneasy may depend on his values, what his values are, and so what has this effect, is still a fact about him.[16]

A third misleading feature of Fulford's chapter is that by its end descriptivism, with which the BST is identified, is saddled with the view that "medicine is nothing *more* than a science."[17] I have never said anything like that. Of course "there is more to medicine than just science" (p. 52), since therapy is replete with values. In fact, the thesis of DI, the paper Fulford is discussing, is that medicine has a dual vocabulary, both value-free theoretical (disease) and value-laden practical (illness). But while there is more to medicine than just science, at least, I have argued, there is science. Medicine has a distinctive theoretical foundation in a value-free science of health and disease.

Choice of Goals

Several critics see the BST as covertly value-laden in choosing goals relative to which dysfunction is disease. Engelhardt (1976b) began this line of criticism, first charging me with "imposing the survival of species as an overriding good" (p. 263). He argued that "one can escape from value judgments in ... defining disease and health only if" one either defines disease as merely atypical function, which is clearly unsatisfactory, or

> accepts species survival as the goal (i.e., so that all judgments about disease and health simply presuppose this and no further value judgments). Otherwise, what counts as disease counts so not because of the designs of nature but because of our goals and expectations. (p. 265)

Engelhardt then attacked my supposed choice of species survival as the goal for its clash with medicine's commitment to "the plight of individual persons" (p. 264). He illustrated this contrast with Margolis's example of the sickle gene, noting that sickle-cell anemia, which is caused by the homozygous-S genotype, is the "price paid" by our species "for the increased survival (health!) of the heterozygote" (p. 264) in present or future malaria-ridden environments. This fact means that

> it may be against our interests in the survival of the species to label as diseases states such as sickle cell disease. Our definition of disease in terms of the distress of individuals may lead to diminishing our species adaptability. ... This is not to say that we make a mistake in deciding that sickle cell disease is a disease... , but simply that we have chosen personal goals over long-range species survival. (p. 266)

Thus Engelhardt's criticism was that the BST made a normative choice of the value of species survival, but that this choice was wrong for medicine, which cares about individual patients.[18] To this I replied that the BST said nothing about species survival (WF named individual survival and reproduction as physiological goals), and also that the BST did not choose that or any other value.

> First, ... it is incorrect that biological functions aim at species survival: the units of selection are the genes of individuals, and therefore biological functions work to preserve the individual and his close kin. Secondly, the definition above incorporates no value judgments; it only states what health (i.e. medical normality) *is,* namely biologically normal part-function. The normal–pathological distinction is a reasonable foundation for medical practice because biological normality is almost always in the interests of the patient. Where this presumption fails, however— as with continuous fertility—other values take precedence over health. Although the value of health is usually important, it is also limited. ... [19]

I would give the same answer to more recent versions of this criticism based on indeterminacy in biological goals. Critics argue that if, as Sommerhoff's theory implies, organismic processes serve multiple goals, then the choice in defining disease of individual survival and reproduction over, say, species survival or ecological equilibrium is a normative choice. Thus Miller Brown:

> How are we to determine those highest-level goals of organisms which lower-level processes function to achieve? ... It may be true that what interests the physiologist is what promotes individual survival and reproduction. But Boorse's account was designed to show that the concept of disease is non-normative. At best, what he has shown is that *given* such a choice of highest-level goals, "function statements will be value-free. ... " But such "empirical matters" are significant only in terms of the goals chosen. What assurance do we have that these *are* the goals of the system whose "species design" we have determined?[20]

Brown's last question is odd, since all the goals I listed are Sommerhoff goals of the system. But his point is clear: The BST makes a normative choice, from a whole class of biological goals, of individual survival and reproduction.

My answer, again, is that there is no choice here—that is simply what disease *is*, as the concept is best reconstructed from medical classifications. The real normative choice is medicine's commitment to combat disease, rather than to promote it, to enrich doctors at any cost, to advance world socialism, preserve planetary ecology or biodiversity, or serve infinitely many other possible goals. Unquestionably medical practice rests on a normative choice to combat disease. But that does not show that the meaning of "disease" rests on a normative choice unless one assumes that the meaning of "disease" is fixed by medical practice. And that is just what writers like Engelhardt, Agich, and Brown do assume. Engelhardt, especially, has consistently treated "disease" as a general category embracing whatever doctors of a given time, place, and culture may see as justifying medical treatment. Admittedly, this decision is to be ruled by certain distinctively medical values which Engelhardt describes, i.e., the commitment to relieve individual suffering, disability, and deformity and to protect individual rational free agency.[21] But this analysis misses a crucial point: the gap between medical treatment and the medical concept of disease. Despite all the time physicians devote to preventing conception, aborting pregnancy, or treating its complications, they never describe fertility or pregnancy *per se* as diseases or pathological conditions.

To me this shows that doctors have a notion of disease that is conceptually independent of medical treatment. But Engelhardt gives no weight to this point, sliding more or less indifferently between "disease" and terms like "medical problem" (1986, p. 168) and "clinical problem" as if they were synonymous.

> The term *clinical problem* underscores the fact that an attempt to give a neutralist, purely descriptive, account of disease fails. Diseases stand out for us as problems to be solved, all else being equal. ... [A] disease is a disease because it is disvalued—in a particular way. It is a clinical problem. One may still draw a distinction, somewhat as Boorse drew, between illness and disease. ... Such does not indicate a contextless disease reality. It acknowledges rather that an individual has a disease, a clinical problem, but is not yet, or is not now, ill. ...
> ...Whether it is athlete's foot, tuberculosis, a deformed nose, or unwanted pregnancy, diseases or clinical problems are not good things to have. (Engelhardt, 1986, pp. 174, 175)

In this passage, Engelhardt seems to hold that clinical problems are diseases, and unwanted pregnancy is a disease or clinical problem; therefore, unwanted pregnancy is a disease.[22] But it isn't, despite having been thoroughly medicalized in every other way. Unwanted pregnancy causes suffering and sometimes real disease; it brings people to doctors and hospitals; physicians treat it, by medical methods like drugs or surgery; and it is sometimes covered by health insurance. Medicine could easily have christened it with some bogus disease name like "dysgravidia," or "gestation adjustment disorder," or "organismic hypernumerosis." But medicine did not, and that is because doctors, in general, know what is disease and what isn't, even if they have trouble reducing their knowledge to a tight definition. (Some philosophers, on the other hand, do not know what disease is and do not seem eager to find out.) In an earlier essay, Engelhardt (1984) had explicitly proposed abandoning the concept of disease altogether as outdated and misleading. In my opinion, as the case of pregnancy illustrates, that is just what he has already done in his writings on disease, including his criticisms of the BST.

In sum, the choice-of-goals point does not make the BST normative unless one presupposes a form of normativism: that the concept of disease is defined by medical practice or medical ethics. That is in no way a necessary premise. Granted, physicians have a special professional commitment to treating disease. Similarly, teachers have a special professional commitment to spreading truth rather than falsehood. But that has no tendency to show that truth is a normative concept defined by educational practice. To choose wood over concrete to build your house with is an evaluative choice, but that does not make the concepts of wood and concrete value-laden. Engelhardt intimates that medicine might have chosen to promote what serves species survival, at the cost of individual pain and death. If it had, on my view that would not have made individual death a form of health. It would simply mean that medicine had decided not, after all, to focus on treating disease, and also that we would have had to be much warier of physicians. In some countries at some times, teachers have chosen or been forced to spread lies. That did not make truth falsehood and falsehood truth, though it did mean one should stay away from teachers.

Having said this much, I should note the possibility that the BST is overprecise in choosing its two biological goals over the rest. The fact is that human physiologists have as yet found no functions clearly serving species survival rather than individual survival and reproduction. That is why one can explain function statements in physiology and medicine as relative to these individual goals. By contrast, medicine has been forced to decide whether to recognize reproduction as an independent goal, since pregnancy and childbirth both lower the efficiency of other functions and greatly increase morbidity and mortality. Here, medicine has plainly chosen to regard reproductive functions as normal; hence we know it is false to say that medicine defines normality only relative to individual survival. But since medicine has never faced an equally clear example of an organ or process sacrificing the individual to species or ecological goals, perhaps the medical disease concept is indeterminate as to goal. If, as some suggest, aging is such a case (a thesis presupposing group selection), then medicine may soon be at a crossroads in its disease concept. And, perhaps, its choice of a way to precisify "disease" will then rest on normative considerations of Engelhardt's sort. But even when a concept is precisified one way rather than another for evaluative reasons, the result can still be a value-free concept: cf. "meter," "degree C," or virtually any unit of measure. To think otherwise is the genetic fallacy.

The BST as Bad Biology

Beyond disputes over the analysis of function, several authors charge the BST with using obsolete or oversimplified biology. The BST is said to be a typological Platonic analysis that ignores species change, intraspecific variability, the environmental relativity of adaptation, and other features of modern population-based evolutionary thought. Strictly speaking, this charge does not threaten the BST. Insofar as the BST fits the medical idea of disease, the charge, if true, would simply prove that medicine is using bad biology. To have shown that fact clearly would be a virtue of the BST. One might then follow Engelhardt (1984) or Hesslow (1993) in abandoning the disease concept. The defects of typological biology threaten the BST only if they reveal its divergence from medical disease classification. Later I will consider that charge, which these authors also lay. But since I am not ready to abandon the disease concept, let me first rebut the accusation of naive or objectionable biology in the BST.

Engelhardt, crediting Margolis, sounded the antitypological theme in 1976 in describing

> what must be the conclusion of any evolutionary understanding of biology: within any population, there will be members possessing various traits enabling more or less successful adaptation to the environment in which the species for the most part, at any particular time, is found. ...
>
> There are no standard environments and, as a result, no standard successful adaptations for members of a species. Views such as Boorse's tend to gloss over the important role of intraspecies variability, while at the same time imposing the survival of species as an overriding good. Given an evolutionary biology and variability within a species (not to mention the variability of environments and therefore variability as to what will be functional or dysfunctional), there is no absolute standard with regard to which one can identify an organism as healthy. ... There is simply no single natural design. ... Rather, there are variations, including aging and even special debilities, which may play their role in the overall survival of a particular species. What will count as successful function in one environment may count as disease in another. (Engelhardt 1976b, pp. 263, 264)

Engelhardt used aging and sickle-cell anemia to illustrate these points, and also homosexuality to charge me with ignoring altruism based on kin selection. He makes the same points in his 1986 discussion without appeal to species survival, and with more on homosexuality and new examples of menopause and skin pigmentation.

> It is not just that Boorse overlooks the role of inclusive fitness in evolutionary accounts of "biological design", he is

unsympathetic to the fact that species may in fact be well adapted because of a balance among various contrasting traits. There may not be a single design, but rather a number of designs. When such is the case, one cannot speak straight-forwardly of either species design or species typicality. ...

To decide what is a problem for medicine, one must make reference to a particular environment and a particular set of goals, so that one can understand whether the individual is well adapted. What will be required for the realization of particular goals will differ from environment to environment. If one is a black living in Trondheim without the avail-ability of exogenous vitamin D, then the possession of highly pigmented skin would put one at a disadvantage with regard to survival. One will have a greater risk, for example, of developing rickets. However, if the environment were to include vitamin D-enriched milk, as is the case in modern circumstances, the individual becomes well adapted. So, too, a Norwegian with lightly pigmented skin and without adequate clothing and protection from the sun will run a markedly increased risk of developing carcinoma of the skin if transported to the tropics. The notion of successful adap-tation is context specific and determined by what one wishes to achieve in a particular context. (Engelhardt, 1986, pp. 168, 169)

A harsher verdict on the BST's biology is by van der Steen (a philosopher of biology) and Thung (henceforth, "vdS/T"). Although they do say that my "description of medicine is implau-sible" in using "a typological way of thinking which is foreign to actual medicine," their main point is that for the same reason the BST is "bad biology" (Van der Steen and Thung, 1988, p. 86), employing "idealizations" that are scientific "nonsense" (p. 90).

Boorse's ... typological approach of species designs would be rejected by almost all biologists. ... [P]hysiology books are not a very good source for information on species designs. The subject belongs primarily to taxonomy, and taxonomists, to whatever school they may belong, nowadays emphatically reject the typological way of thinking. ... Philosophers concentrating on taxonomy share their opinion. For example, David Hull ... wrote a paper in 1965 with the ominous title "The effect of essentialism on taxonomy—two thousand years of stasis" ("essentialism," in the present context, roughly means "typology"). One of his points is that species names cannot be defined in terms of necessary and sufficient properties. ... Variability within species simply does not allow that, even if we concentrate on species here and now and, like Boorse, disregard evolutionary time-scales. ...

Variability within species is also an important theme in population genetics. ... (p. 88)

VdS/T's population example is, again, the sickle gene. I could handle this case, they say, by calling anyone with that gene abnormal, but then there is so much similar variation in populations that everyone will be abnormal. "[W]hy," they ask, "should one make ideals which biology rejects a paradigm?" After objecting to my contrast of internal vs environmental causation, they end by claiming that the BST is idealized even in physiology.

Boorse's conception of design becomes inadequate even within physiology, his favourite discipline, as one leaves the realm of elementary textbooks of medical physiology. Physiological functions change with the environment, so *there are no reference values simpliciter.* Reference-values will have to be *context-dependent.* They are sensible only if they are related to the environment besides age and sex. Blood cell counts change with altitude, metabolic rates with temperature, and so on. Clinical diagnosis in fact makes allowance for such items. Many conditions which are pathological according to Boorse's definitions may be biologically normal in some, and abnormal in other environments. Plain biology would not justify their being classified as diseases in our nomenclatures.

Boorse apparently wants to use idealizations representing optimal designs in hypothetical environments without pathogens and harsh physical conditions. Such environments

are not very "biological". They represent ideals one would
like to realize. Boorse's idealizations are nonsense in the
context of biology. But they may have a function if we
regard them as a means for expressing human values.
On that interpretation Boorse is just another normativist.
(p. 90)

Mon Dieu, quelle insulte! But is the BST's biology really as
bad as vdS/T say? First, what did it say on these issues?

In HTC, I noted that the idea that the normal is the natural
makes health and disease members of "a family of typological
and teleological notions which are usually associated with Aristotelian biology and viewed with suspicion" (p. 554). But just as
recent philosophy had found "aseptic substitutes" for function
and goal-directedness, so the BST was to be a similar aseptic
substitute for the idea of diseases as foreign to the nature of the
species. To block all ancient infection, I described species design
as a "statistical abstraction" (p. 558) empirically derived from the
population, and I allowed for population variability in two ways:
the central role of the statistical-subnormality concept (Fig. 1), and
the point that polymorphic forms (blood type or eye color) can be
equally normal. On such an "empirical ideal" I further commented:

> It would be a mistake to think that this notion of a spe-
> cies design is inconsistent with evolutionary biology,
> which emphasizes constant variation. The typical result
> of evolution is precisely a trait's becoming established
> in a species, only rarely showing major variations under
> individual inheritance and environment. On all but evo-
> lutionary time scales, biological designs have a massive
> constancy vigorously maintained by normalizing selec-
> tion. It is this short-term constancy on which the theory
> and practice of medicine rely. (p. 557)

Evidently Engelhardt and vdS/T believe that some Hellenic
sepsis remains in this analysis, but where? Despite consulting
sources vdS/T recommended, I cannot find a false biological
statement in HTC, nor do vdS/T cite any specific one. Rather,

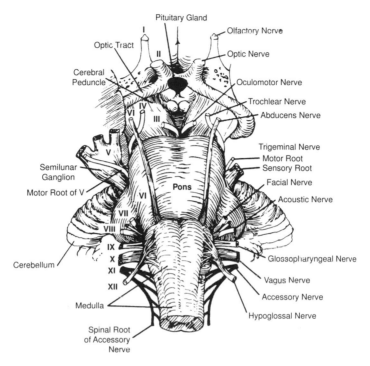

Fig. 3. The twelve pairs of cranial nerves. Figure adapted from *Correlative Neuroanatomy and Functional Neurology,* 18th edition, by J. G. Chusid, Lange Medical Publications, Los Altos, CA 1982.

these authors' complaints about the BST's biology seem to have two possible interpretations.

One possible thesis is that, empirically, variability is so extreme in our species and others that no significant "species design" can be described, or none detailed enough to be a theoretical foundation for medicine. This seems to me to amount to saying that human-physiology (or insect-physiology) textbooks are actually false, i.e., that the endless array of facts about human physiology that medical students learn in their first or second year is untrue. I invite critics pursuing this claim to do so in detail about some moderately complex physiological topic—the functional architecture of the kidney, say, or the hepatic vasculature, or the twelve cranial nerves (Fig. 3, Table 1). Surely vdS/T and

Table 1
Overview of Cranial Nerves

| | Functional type | Functions | | | Location of cell bodies | | Major connections |
		Motor innervation	Sensory function	Parasympathetic function	Within sensory organ or ganglia	Within brain stem	
Special sensory							
I Olfactory	SS		Sense of smell		Olfactory mucosa		Mucosa projects to olfactory bulb
II Optic	SS		Visual input from eye		Ganglion cells in retina		Projects to lateral geniculate; superior colliculus
VIII Vestibulocochlear	SS		Auditory and vestibular input from inner ear		Cochlear ganglion		Projects to cochlear nuclei, then inferior colliculi, medial geniculate
					Vestibular ganglion		Projects to vestibular nuclei
Motor for ocular system							
III Oculomotor	SE	Medial rectus, superior rectus, inferior rectus, inferior oblique				Oculomotor nucleus	Receives input from lateral gaze center (paramedial pontine reticular formation; PPRF) via median longitudinal fasciculus
	VE			Constriction of pupil		Edinger-Westphal nucleus	Projects to cillary ganglia, then to pupil
IV Trochlear	SE	Superior oblique				Trochlear nucl.	
VI Abducens	SE	Lateral rectus				Abducens nucl.	Receives input from PPRF

Category	Nerve	Fiber	Function	Ganglion	Nucleus	Projection
Other pure motor	XI Accessory	BE	Sternocleidomastoid, trapezius		Ventral horns at C2–C5	
	XII Hypo-glossal	SE	Muscles of tongue, hyoid bone		Hypoglossal nucleus	
Mixed	V Trigeminal	SA	Sensation from face, cornea, teeth, gum, palate. General sensation from anterior two-thirds of tongue	Semilunar (= gasserian or trigeminal) ganglia		Projects to sensory nucl. and spinal tract of V, then to thalamus (VPM)
		BE	Chewing muscles		Motor nucl. of V	
	VII Facial	BE	Muscles of facial expression, platysma, stapedius		Facial nucl.	
		VA	Taste, anterior two-thirds of tongue (via chorda tympani)	Geniculate ganglion		Projects to solitary tract and nucleus, then to thalamus (VPM)
			Submandibular, sublingual, lacrimal glands (via nervus intermedius)		Superior salivatory nucleus	
	IX Glossopha-ryngeal	VE	Parotid gland		Inferior salivatory nucl.	

(Continued)

Table 1 (*Continued*)

	Functional type	Functions			Location of cell bodies		
		Motor innervation	Sensory function	Parasympathetic function	Within sensory organ or ganglia	Within brain stem	Major connections
	VA		General sensation from posterior one-third of tongue, soft palate auditory tube. Sensory input from carotid bodies and sinus. Taste from posterior one-third of tongue		Inferior (petrosal) and superior glossopharyngeal ganglia		Projects to solitary tract and nucleus
	BE	Stylopharyngeus muscle				Ambiguous nucl.	
X Vagus	BE	Soft palate and pharynx				Ambiguous nucl.	
	VE			Autonomic control of thoracic and abdominal viscera		Dorsal motor nucleus	
	SA		External auditory meatus		Superior (Jugular) ganglion		Projects to thalamus (VPM)
	VA		Sensation from abdominal and thoracic viscera		Inferior vagal (nodose) and superior ganglia		Projects to solitary tract and nucleus

Efferent (motor): SE—somatic, general SE; BE—branchial, special VE; VE—visceral, general VE. Afferent (sensory): VA—visceral, general VA; special VA; SA—somatic, general SA; SS—sensory. Most nerves with SE components have a few SA fibers for proprioception. Table adapted from *Correlative Neuroanatomy*, 22nd edition, S. G. Waxman and J. DeGroot, eds., Appleton and Lange, Norwalk, CT, 1995.

Engelhardt are not suggesting that the kidneys of some human groups do not contain cortex, medulla, and pelvis, with multiple functional units (nephrons) including a loop of Henle operating by the countercurrent-multiplier principle. Surely their point is not that we should expect some human kidneys to work instead by carbon filtration, or magnetic resonance; that there are normal humans with only four cranial nerves, or 17, or 33; that the vagus nerve (#10) is sometimes #5, sometimes #8, sometimes #31, and sometimes absent altogether, with all nonhormonal visceral cholinergic stimulation carried by tiny fiber-optic cables or positron beams.

Perhaps unintentionally, these authors' writings leave the impression that human variability is so great that "species design" can have no real content. To think this is to deny the obvious, to err about what Mayr calls the "order of magnitude" (Mayr, 1988, p. 329) of morphologic difference. Actually, one could easily pursue the opposite theme from vdS/T and Engelhardt's intraspecific variation: the massive overlap of structure between species. In a different essay Mayr notes that

> in most higher taxa there is an extraordinary basic similarity of all members. This is why a good zoologist can usually tell of any animal not only to what phylum but often also to what class, order, or even family it belongs. (Mayr, 1988, p. 109)

If one were to go by DNA overlap, well over 90% of the human design is primate design, not specifically human. Of that, the vast majority is mammalian, not specifically primate; of that, most is vertebrate, and so on. Indeed, considerable structure and function is shared by all organisms on earth. So I continue to think it is fair to say:

> Our species and others are in fact highly uniform in structure and function; otherwise there would be no point to the extreme detail in textbooks of human physiology. This uniformity of functional organization I call the species design.

To deny its existence on Darwinian grounds would be to
miss the forest for the trees. (HTC, p. 557)

I think that authors like Engelhardt and vdS/T do miss the
forest for the trees—e.g., if heterosis so pervades human physiology, why does every author use the same example, the sickle
gene? But they have other complaints, one of which is still pure
biology. Clearly, what "almost all biologists," or taxonomists,
believe is not that medical textbooks are tissues of lies, or even
wildly misleading. Rather, what almost all biologists reject is
some kind of typological thinking, and, in particular, "essentialism." VdS/T lay a general charge of typology against the BST.
However, if we follow Sober's analysis, essentialism (the topic
of the Hull paper that vdS/T cite) is only one component of the
complex of typological ideas in Aristotelian biology. Sober's
long discussion (1984, pp. 155–169; 1980, reprinted in 1994)
finds the following major ideas in Aristotle: that all objects in
nature have a natural state, which they will be in unless acted on
by interfering forces (the "natural state model"); that the most
natural act of any organism is reproduction, which, apart from
interfering forces, will produce identical offspring, so that all
intraspecific variation is unnatural; that there is therefore a natural course of individual development toward a specific phenotype; and that every species has a causal essence, possessed by
every member, that tends to cause the species' natural form. As
to the "natural state model," Sober notes that it is retained by both
modern physics (rest vs force) and evolutionary biology, though
the latter reverses Aristotle by taking variation as a natural state
disturbed by selection. What he believes contemporary biology
has decisively rejected is essentialism, which he describes as a
"doctrine about causal mechanisms." That is, "essences are necessary properties that play a certain causal (and hence explanatory) role" (1984, pp. 164, 165).

So understood, I fail to see how the BST emits even a whiff
of essentialism. It has little to say about causal powers, and nothing about defining species by necessary properties, the

taxonomist's version of essentialism. Essentialism is not the kind of typology I meant to concede at all. Rather, I meant that the BST is, for example, more typological than Darwin, since it relativizes health to species, whereas he considered a species to be as arbitrary a taxon as a subspecies. But in this the BST better fits contemporary biology, which regards species (at least as defined by Mayr's "biological species concept") as real and crucial units with describable characteristics.

> Modern biologists are almost unanimously agreed that there are real discontinuities in organic nature, which delimit natural entities that are designated as species. Therefore the species is one of the basic foundations of almost all biological disciplines. Each species has different biological characteristics, and the analysis and comparison of these differences is a prerequisite for all other research in ecology, behavioral biology, comparative morphology and physiology, molecular biology, and indeed all branches of biology. Whether he realizes it or not, every biologist works with species. (Mayr, 1988, p. 331)

My "species design" is nothing but the functional component of Mayr's "biological characteristics." Mayr does say that the "typological species concept" is no longer useful, because of sibling species and polymorphism. But the BST can accommodate normal polymorphisms, and, of course, admits normal statistical variation both between and within individuals.[23]

Indeed, by Mayr's own characterization of typology vs population thinking—Mayr being essentialism's "sternest critic" (Rosenberg 1985, p. 191)—the BST is indisputably populationist. Mayr says the essentialist holds that there are

> a limited number of fixed, unchangeable "ideas" underlying the observed variability, with the *eidos* (idea) being the only thing that is fixed and real, while the observed variability has no more reality than the shadows of an object on a cave wall. ... [By contrast,] the populationist stresses the uniqueness of

everything in the organic world. ... All organisms and organic phenomena are composed of unique features and can be described collectively only in statistical terms. Individuals ... form populations of which we can determine only the arithmetic mean and the statistics of variation. Averages are merely statistical abstractions; only the individuals of which the populations are composed have reality. The ultimate conclusions of the population thinker and of the typologist are precisely the opposite. For the typologist, the type (*eidos*) is real and the variation an illusion, while for the populationist the type (average) is an abstraction and only the variation is real. No two ways of looking at nature could be more different. (Mayr, 1976, pp. 27, 28)

Apart from Mayr's remarks about "reality," which Sober criticizes (1984, pp. 157, 166), my model of health in HTC took care to fit his populationist position. For HTC as for Mayr, a species, though a nonarbitrary grouping,[24] is a variable population of unique individuals for whom a statistical abstraction of normal functioning defines health. In sum, vdS/T conflate all typology with essentialism, and seem too concerned with my efforts to show continuity between ancient and modern disease concepts, and too little with my efforts to distinguish them. An "aseptic substitute" is, after all, a substitute.

One might, of course, claim that the HTC is unacceptably typological simply in calling some members of a population abnormal—i.e., that the concept of normality is unbiological *per se*. But Sober and many other biologists and philosophers of biology freely use medical terms like "disease," "pathogen," and so on in discussing such phenomena as myxomatosis in Australian rabbits.[25] Indeed, one of the BST's advantages over its rivals is that it explains pure biologists' application of disease language to wild animals, as well as the usage of veterinary medicine and plant pathology.[26] If these authors have some nonmedical idea of disease, pathology, and so on that does not use the concept of normality, it would be helpful for them to offer it.

In any case, that normality itself is an unbiological concept is not the position of Engelhardt and vdS/T.[27] They claim, on the contrary, that medicine uses differential adaptation as a criterion of disease. My view is that this is simply false: that is just what medicine does not do. Being black in Trondheim or white in the Sahara is neither a disease nor pathological, though each condition raises the risk of disease, as Engelhardt observes. His examples, far from being cases where maladaptation is disease, are instead clear cases of the contrast between health and adaptation. Since Engelhardt often stresses ways in which the BST looks unsympathetic to homosexuals and women, it is ironic to find him harking back to Benjamin Rush in implying that Negritude is a disease in Scandinavia.[28] One can avoid that conclusion very simply by distinguishing the pathological from the pathogenic.

The BST as Bad Medicine

General Features

Broad vs Narrow Senses of "Disease"

As many writers note, even in medicine "disease" is mostly used more narrowly than by the BST. It is not usual to call "diseases" such conditions as injuries (broken arm, dog bite), deformities (club foot, cleft palate), static abnormalities (foreign bodies in the stomach), functional impairments (blindness, deafness), poisonings (arsenic or alcohol intoxication), environmental effects (sunburn, heatstroke, frostbite), and various other phenomena (starvation, drowning). It is also usual to distinguish among diseases, signs, symptoms, and sometimes syndromes as well. Some writers see in this point an objection to the BST. Continuing Whitbeck's original charge (1978), Miller Brown says the BST "includes too much," and stresses that both the AMA *Standard Nomenclature* and the WHO *International Classification of Disease* "distinguish among 'diseases, injuries, impairments, and simple anomalies'" (1985, p. 317). Reznek, in his long and careful discussion of various kinds of "negative medical conditions," agrees with Whitbeck that diseases must at least be processes, not states (Reznek, 1987, p. 72). But he correctly recognizes that "pathological condition" is a broad medical term encompassing all the types of medical abnormalities he and others discuss.

This whole issue is, to me, a red herring. The BST's target was the medical concept of health. Health is constantly described by medical sources both as "absence of disease" and as normality in the sense of absence of pathological conditions. Both these statements are medical clichés, hardly felt to be controversial at all. But insofar as medicine employs any such negative concept of health, narrower uses of "disease" are irrelevant to the task of analyzing it. If a blind, paraplegic, diabetic, starving dwarf who has eaten only arsenic for weeks is dragged into the Arctic snow by a pack of vicious wolves who tear pieces off his frostbitten body until he is crushed by a tree in an avalanche, it would be absurd to say that he was perfectly healthy from beginning to end. No one can be healthy and dead, but one can die without a disease in almost any of the narrower senses of disease. Consequently those who, like Whitbeck, Brown, and Reznek, claim that the broad usage of "disease" is incorrect are claiming that health has no chance whatever of being the absence of disease,[29] and so that all the medical sources using this formula are in error. My approach was more linguistically tolerant. However, if someone wants to reject the broad usage of "disease" completely, it is fine with me as long as the term "pathological" remains. The point is that medicine does employ a negative concept of health as the absence of any of what Reznek calls "negative medical conditions," and the most usual omnibus term is, in fact, "pathological," with "morbid" a close second, especially in the British Commonwealth.

A secondary problem is, of course, to analyze the narrower sense or senses of "disease" themselves. I have said nothing on this topic, except that I doubt the problem is of much interest because

> [n]arrower usages of 'disease' are seldom clear or consistent in medical writings: Acosta's disease and caisson disease are

environmental effects, the disease hemosiderosis is chronic iron poisoning, and even classic infectious diseases like diphtheria resemble injuries and poisonings in various ways. (CH 363, 364).

One who thinks there is some important fixed narrow content to "disease" should test his hypothesis against a dictionary of eponymous diseases ("X's disease"). For discussion of many noneponymous examples, *see* Reznek (1987, pp. 65–73), who comes to the same conclusion I did: "There does not seem to be some property that diseases have and that other negative medical conditions lack" (p. 73). That is, as Reznek says, disease in the narrow sense lacks both a real and a nominal essence.

The main importance of the broad/narrow distinction seems to me to be simply that one must take care not to be led off an analytic track by narrow senses. That is what happens, I believe, to Hare and also to Culver, Gert, and Clouser (CGC) in their article (1981) and book (1982). Starting from a narrow sense of "disease," CGC noted the need for a "genus term" covering the diverse medical categories of diseases, disorders, illnesses, injuries, defects, and so forth. Proposing the term "malady" for this role, they offered a definition of malady. Martin (1985) then argued that by CGC's definition menstruation, menopause, and pregnancy are all maladies and accused the authors of discrimination against women. In reply to Martin, CGC (1986) conceded that pregnancy is a malady and tried to justify this decision and dodge the bias charge.

This concession shows that whatever CGC had been doing— and it was always a bit unclear what that was (Martin 1985, p. 331), since it is unclear what constraints authors face in defining their own term for an undetermined extension—they were not, as their title might suggest, offering an analysis of "disease" in any sense, narrow or broad, or of "pathological." If we are explaining negative theoretical health in medicine, when a definition is found to classify pregnancy as disease, pathological, abnormal, or anything of the sort, that is an intolerable result. No analysis faithful to medicine can count pregnancy as *per se* unhealthy. By contrast, the BST's problems in fitting medicine seem fairly minor. Medicine's conceptual scheme would hardly collapse if it had to stop calling macacus ear or weird hymens pathological, or even to admit that there are no universal diseases at all, only premature aging, excessive atherosclerosis, and so on. But to call pregnancy *per se* unhealthy would strike at the very heart of medical thought; it is the analytic equivalent of the "Game Over" sign in a video game. The fact that CGC accept pregnancy as a malady shows that whatever they are doing, they cannot be offering absence of malady as an analysis of health in medicine.[30] Their initial error, I think, was not to see that medicine already has a "genus term" for conditions inconsistent with theoretical health, viz, either "disease" in the broad sense or "pathological."

Theoretical/Practical, Pathological/Clinical, Disease/Illness

Many writers attack the BST either by attacking its distinction between theoretical and practical health—between pathological and clinical levels of description—or by assigning "disease" to the clinical level.

Agich criticizes the BST for taking an "abstract philosophical approach to medicine" (Agich, 1983, p. 28), when medicine is essentially a concrete practice embedded in certain social institutions. Thus, even when medicine speaks of "disease" it employs values:

> To call a state of affairs a disease, then, is not simply to describe it in species typical or biological terms, but to characterize it as somehow bad or undesirable relative to human freedom in general and various particular values. Such judgments can and do conflict with other judgments regarding an individual's experienced sense of illness or well-being. This conflict is eminently practical and social. (Agich, p. 36)

Agich cites Engelhardt's inventory of medical values as "algesic (pain related), aesthetic, and teleological" (p. 36). In general, his position seems consistent with Engelhardt's, though I cannot tell whether Agich too is content not to distinguish between diseases and medical or clinical problems.

Interestingly, in arguing that "disease," as well as "illness," is a practical category, Agich cites two writers who seem to take opposite positions on the same example, athletic ability. A brief passage from Robert Brown is first quoted, criticizing me for being unable to answer the question "Unnatural deviation from what adequate and desirable level of performance?" so as to separate disease from illness. Brown had claimed that disease and illness must both be defined in terms of "incapacitating disturbance in function or structure" (Brown 1977, p. 27), and therefore that my distinction between the two evaporates. One of Brown's examples of the problem of performance standards had been athletics: "Healthy blood pressure for a young Olympic athlete is very different from that which can be tolerated by a sixty-year-old housewife" (p. 21). Since this example occurs in Brown's section "Recognizing disease," I take it he was suggesting that the housewife's performance level might be a disease in the athlete, and in general, that the same condition can be a disease or not depending on the individual's life situation, activities, and goals. Erde, whom Agich also quotes, seems to make exactly the same point, the dependence of health on what individuals want to do. But Erde curiously says just the opposite about the athlete, that an injured athlete who is not yet healed enough for competition is not thereby diseased (Erde, 1979, pp. 42, 43).

The short answer to all these writers is that they confuse theoretical and practical, pathological and clinical, description. On the theoretical level where pathologists operate, it is false that pathology depends on what a person wants or should want to do, how he views a condition, or his life situation. Obviously, such factors determine the clinical or social importance of disease states—how much to care about them, how far to investigate them, what treatment to give them, whether insurance should pay for them—but they do not affect what is theoretically a disease in the first place. Whether a condition is pathological does not depend, as Brown suggests, on what a specific patient's career or lifestyle can "tolerate." Many people's lifestyle can tolerate huge amounts of disease, while, conversely, athletes will find many normal conditions intolerable, such as average strength, speed, or coordination. Brown also asks about Richman's example (1975, p. 426) of a nonharmful disease, a treatable infectious skin discoloration: "what makes this condition a disease or disorder if it does not affect the agent's normal functions?" (p. 25). I answered a similar example from Engelhardt in HTC: the infection involves local dysfunction of skin cells. To the pathologist, any process causing cellular dysfunction, no matter how local, is pathological. But this judgment has nothing to do with how an individual patient views it or how much it affects his life activities. Although even theoretical disease judgments seem relative to age, it is, I think, just false that in two men of the same age, the same physiological condition could be pathological in one because he is an athlete and normal in the other because he is not.[31]

We must note that Engelhardt, Agich, Brown, Margolis, Fulford, and many other writers are following what Fulford calls the "reverse-view" analytic strategy: to start with illness and define disease in terms of its causal relation thereto. In this strategy, some explanation is first given of the "molar" (Margolis, 1976, p. 246) concept of illness, in terms of the patient's overt distress, disability, and so on. Then disease is defined by its tendency to cause illness. Disease may be said, for example, to be a scientific category for explaining and predicting illness, competing with other explanations (Engelhardt, 1976b). Illness, in this approach, is often seen as "reflexively palpable disease": "Illness is simply a diseased state manifest to an agent through that agent's symptoms. ... "[32] Disease is broader than illness because it is whatever tends to cause illness: "the presence of causal factors that are likely to produce illness" (Margolis, 1976, pp. 241, 242) or those "physiologically, anatomically, or psychologically rooted circumstances" that "surely," "likely," or "usually lead to suffering" (Engelhardt, 1986, p. 174).

Unfortunately, I believe, every one of these ideas is wrong. It is at least dubious that the field of illness can first be defined without reference to disease or pathology, i.e., to internal dysfunction. Many conditions cause suffering or disability, such as disappointment in love or a lack of violin talent, and may also bring people to doctors, e.g., ugliness or unwanted pregnancy, yet are not pathological. To me, that means they are not genuine sickness or illness, though this does not entail that doctors cannot or should not help with them. Even if we could phenomenologically characterize some primitive illness experience, it would still be false that illness is disease evident to the patient, since myopia and athlete's foot, though quite evident, are not illnesses. Conversely, one can be sick without knowing it, a psychiatrically crucial point. The fact is that all of us got illness wrong: the key concept is systemic incapacitation.[33]

As for the causal relation of disease to illness however defined, the true relation is at best extraordinarily weak, far weaker than these authors' formulations. It is not true that all disease or pathology is of a type "likely" to cause clinical illness. Quite often, most cases of a disease are subclinical or "lanthanic," in Feinstein's terminology (1967, p. 145). Pathologists even report malignant growths most cases of which are incidental autopsy findings, e.g., carcinoids and one type of prostate cancer. To be at all plausible, efforts to define disease via illness would have to be much weaker. For example, one might say that disease is anything that raises the probability of illness, or anything that can cause illness. But each of these is too broad unless at least further restricted to the species-atypical, since the appendix and various human design defects, like the normal concentration of bile or the crossing of the respiratory and alimentary systems, fit these descriptions. Still worse, since cellular pathology can be very local—pathologists constantly talk of focal necrosis, pinpoint hemorrhages, and the like—many types of pathology will never cause illness unless aggregated.[34] Yet aggregating almost any normal substance or process, even blood oxygen or erythrocytes, heartbeat or respiration, can also cause illness—cf. such pathological states as polycythemia, hyperthyroidism, ventricular fibrillation, and so on.

The true analytic relation between molar and molecular, or "macro" and "micro," concepts in biomedicine is that the organism's gross output shows the goals against which the fine-structure of part-functions is defined. But just as physiologists trace this fine-structure down to tissue, cell, organelle, and genetic levels, so pathologists also recognize microdysfunctions as pathological. As both Wakefield and I have stressed, a biological function of a trait need only make its bearers marginally more likely to survive and reproduce. However, an effect manifest on evolutionary time scales may be all but invisible in an individual lifetime.[35] If one further recognizes the overcapacity and redundancy in biological designs—large livers, two kidneys, multiple biochemical pathways and regulatory systems—one can see that any necessary relation of local pathology to overt illness is so weak as to be almost useless for defining disease.

At bottom, disease is a pathological, not a clinical, concept, in that all sorts of subclinical pathology can exist without, or before, clinical manifestations. One can, of course, define a concept of "clinical disease": disease that is clinically detectable or treatable, or what I call a diagnostic or therapeutic abnormality. But one cannot go in the opposite direction, defining disease from clinical concepts.[36] Wakefield's paper illustrates that this point and the pathological/clinical contrast are no small matters. Wakefield sets out to define "disorder" for purposes of psychiatric classifications. But like Spitzer and other participants in the DSM-III project, Wakefield means disorder to be a clinical concept. Spitzer and Endicott, for instance, stated that the concept of disorder includes an implicit "call to action" (Spitzer and Endicott, 1978, p. 17) by health–care personnel. Accordingly, both Wakefield and Spitzer ignore the distinction between pathological and clinical concepts, and that is why Wakefield feels he must supplement dysfunction with a harm clause:

> To be considered a disorder, the dysfunction must also cause
> significant harm to the person. ... For example, a dysfunc-
> tion in one kidney often has no effect on the overall well-
> being of a person and so is not considered to be a disorder;
> physicians will remove a kidney from a live donor ... with
> no sense that they are causing a disorder, even though people
> are certainly naturally designed to have two kidneys. (pp.
> 383, 384)

If "disorder" here is to mean disease in the broad sense, or
pathological condition, this quotation is nonsense. Of course
unilateral kidney disease is disease. Of course surgically removing a kidney creates a pathological condition, unless one falsely
assumes that doctors never properly cause pathology. Since
Wakefield's theory (disorder = harmful dysfunction) includes
the biological requirement of dysfunction, perhaps no great damage is done. What he is analyzing is really "clinical disease"
(more exactly, "therapeutic abnormality"), with dysfunction analyzing the "disease" part and harm analyzing the "clinical" part.
But surely it is preferable for psychiatry, like medicine, to distinguish between pathological and clinical levels, between psychopathology *per se* and clinical psychopathology important enough
to need treatment. Confusing theoretical and practical issues has
already impeded debate over conditions like homosexuality, as I
noted in DI and CH. It is also unfortunate that Wakefield merges
therapeutic into social values (Wakefield, 1992, p. 373), since
purely medical (diagnostic and therapeutic) health categories can
surely be usefully divided from social ones (disability, insanity).
In general, it is hopeless to expect a single concept of "disorder"
simultaneously to fulfill theoretical, clinical, and social roles.
DSM-III can ignore the theoretical/clinical contrast partly because
it is intentionally atheoretic. But its atheoreticity is much criticized by Wakefield and others, and "the history of medicine shows
that an atheoretic, purely clinical approach to identifying diseases is of limited value" (CH, 382). By contrast, I begin with
theoretical abnormality (roughly Wakefield's dysfunction) and

would then define clinical normality or social normality by adding evaluative tests to that basis. Only the BST, I think, in this way has a chance of matching the richness of our health vocabulary in and out of medicine.

Finally, it is the fact that "disease" is basically a pathologist's concept which answers the concerns of several authors that the BST is too demanding. Wulff et al. (1986, p. 48) observe that by purely statistical criteria, the normal range for a laboratory test is chosen to include 95% of the "healthy" population; then any patient who is given 25 lab tests has a 72% chance of being found abnormal. Thus, as Murphy quipped, "a normal person is anyone who has not been sufficiently investigated" (Murphy, 1976, p. 123). But first, insofar as the 5% rate for each test is an error rate, then the 72% is only a probability that at least one test is wrong; that is an epistemic, not a conceptual, problem. Second, many lab tests are not direct tests of function, but of some quantity correlated with function, and so once again relate to detecting, not defining, abnormality. Third, the calculation assumes the tests are independent, but for genuine function tests they may not be since they may correlate with general health status. But fourth, it is certain in any case that to the pathologist no one is normal. Murphy's remark is literally, uncontroversially true on the pathological level of description, since quite apart from disease in a narrow sense, everyone has at least assorted scars external and internal, and almost everyone has some current injury, if only a tiny bruise, muscle tear, or shaving cut.

Thus, even if the ubiquity of heterosis forced the BST to count everyone as diseased, as vdS/T suggest in discussing sickle trait, that is no objection. Everyone is already diseased regardless of how we treat heterozygote superiority. I give the same answer to Nordenfelt's one-dead-cell objection (Nordenfelt, 1987, p. 28). At the pathologist's level of description, there is no paradox in calling one dead cell pathological, except, of course, in tissues like skin and mucosa whose normal function entails constant death and regeneration. One dead cell is just the ultimate in focal necrosis, one of pathology's most common findings. Why would one dead neuron not be pathological? Is a dead neuron a normal neuron? Do dead neurons function normally? Since the neuron is a body part, in a tissue whose physiological function does not feature regular regeneration, it is not much more mysterious that its death is pathological than that the whole organism's death is pathological. (For a related reply to Nordenfelt *see* Defense Mechanisms.) Of course, one dead neuron is a trivial piece of pathology. But to call a condition pathological implies nothing about its importance. To think otherwise is to confuse theoretical and clinical normality. Judgments of importance belong to clinical practice, not to basic theory of normality. Unfortunately most authors, even Reznek, try to get along with only one disease-concept to cover theoretical, clinical, and even social domains, which is plainly impossible.

Medical Theory as Nonexistent, Inadequate, or Value-Laden

Critics charge that "disease" cannot be a value-free scientific concept of medical theory because:
1. Medicine is not science;
2. "Disease" is theoretically useless; or
3. Science in general, or biology or medicine in particular, is value-laden.

No Medical Theory

Margolis sounded a theme that resonated with several other authors.

> [M]edicine is primarily an art and, dependently, a science: it is primarily an *institutionalized service concerned with the care and cure of the ill and the control of disease...* (1976, p. 242)

If there is no medical science, then *a fortiori* there is no theoretical medical science, which is the view of Miller Brown.

> There can be, then, only in a very indirect and derivative sense a *theoretical* medicine, and accordingly it makes little

sense to argue that 'disease' is a theoretical concept of medicine. Medicine is largely a practical activity since it is concerned with diagnosis and treatment of illness, usually as a
matter of medical crisis; and, as such, it is more akin to a
technology. Theory *in* medicine, as in electronics, is bor-
rowed from fundamental sciences like biology, chemistry,
and physics. Research in medicine, when it is not biology,
chemistry, and physics, is a kind of technological enterprise
allied to these sciences and only rarely leading directly to
development in theory.[37]

Not surprisingly, Brown's paper appeared in *Journal of Medicine and Philosophy,* not its rival *Theoretical Medicine.*

This objection seems to be of that rare variety which can be
answered in one word: "pathology." Pathology, one of the "basic
sciences" studied in the first or second year of medical school, is
the scientific study of disease. It is based on anatomy, physiology, biochemistry, genetics, and other biological sciences, but is
distinctively medical in being wholly devoted to disease.[38] It
comprises whatever general principles about disease can be stated,
plus descriptions of basic manifestations and recurrent types of
pathologic reaction with whatever generality is possible, plus
specific disease entities. Since it certainly uses the term "disease," it is the venue of a scientific, theoretical, medical disease
concept unless either pathology is not science, or not theoretical,
or not medicine. To call it unscientific would be odd, since, of the
various demarcation tests philosophers of science have proposed,
pathology seems to pass every one that biology passes. That
pathology offers no genuine theory of disease is Hessler's point
(2). That leaves the claim that if pathology is a theoretical science, it is biology, not medicine. I do not know why one must in
this way read pathology out of medicine, as Engelhardt, Margolis, Brown, and others seem to do. Brown mentions "research
into disease etiology, pathology, and treatment" (Brown, 1985, p.
326). Does he realize that there is an actual discipline called
"pathology" that counts as one of the basic sciences, is housed

almost entirely in medical schools, and so is, one would think, part of medicine and a suitable venue for medical theory?

Even if one declares pathology part of biology, it is still a scientific discipline concerned with disease. Many biologists talk of diseases, pathogens, epidemics, and so on in any event. Thus, whether pathology is biology or medicine, the BST can claim that it furnishes medicine with a basic scientific concept of disease as a scaffold on which medicine, and society at large, can build clinical and social disease concepts. Agich (1983, p. 29) complains that my analytic methodology was unclear. That seems unfair, since I was the first philosopher (in HTC, which he does not cite) to emphasize the importance of matching the stock of diseases recognized by medical usage. At any rate, I can make my methodology even clearer. I am content for the BST to live or die by the considered usage of pathologists—which does not, of course, exclude that on reflection (as in Rawlsian equilibrium), pathologists might revise their usage slightly to achieve consistency with a simple and powerful theory.[39]

"DISEASE" THEORETICALLY USELESS

The general thesis of Germund Hesslow's stimulating paper (1993) is that

> the crucial role of the 'disease' concept is illusory. The health/ disease distinction is irrelevant for most decisions and represents a conceptual straightjacket. (p. 1)

Hesslow endorses something like the BST as the best possible analysis of the medical disease concept. But he finds that concept theoretically useless and practically irrelevant to both clinical treatment and social issues like medical insurance, institutional leave, and moral and legal responsibility. Actually, I agree with almost all of Hesslow's observations, but not his conclusions from them. I have myself made some of his points about the independence of theoretical disease from practical judgments of all kinds (DI, MH, CH). I am simply not so pessimistic as he about the continued value of health concepts. One should note that, despite Hesslow's title, he applies his thesis to "health" just as much as to "disease."

Regarding theory, after canvassing several motives for clear definitions in science, Hesslow concludes that none of these motives applies to "disease" because it is "not a theoretical term in any scientific theory":

> There is no biomedical theory in which disease appears as a theoretical entity and there are no laws or generalizations linking disease to other important variables. (p. 5)

I have always suspected that Hesslow is right on this point, and so titled "Health as a theoretical concept" with some trepidation. Even in pathology, medicine's prime theoretical arena, one looks in vain for any disease counterpart to Kepler's, Newton's, Einstein's, or Maxwell's equations. Of course, counterparts to physics are thin on the ground even in biology. But it is still hard to find anything purely medical corresponding to the Hardy-Weinberg law, or to Mary Williams' axiomatic version of Darwinian theory.

But even if "disease" is, as Hesslow says, "not a theoretical term in the same sense as ... 'electron,' 'force,' or 'gene'" (p. 10), it still seems to me useful for what he calls "intellectually organizing a certain body of knowledge" (p. 4). If disease is only as scientifically useful a concept as learning, memory, and intelligence are in psychology—his examples—that is still useful indeed. To a degree, Hesslow's skepticism may be trading on a contrast between scientific domains where theory has made more progress (physics, genetics) and less progress (psychology, pathology). But since he appears to take the same line on many basic biological concepts as he does on disease, that there is no point in conceptual analysis, perhaps I can simply grant everything he says about theory. Certainly, I am satisfied if "disease" is as much a theoretical term (and as value-free!) as "animal," "plant," "cell," and "hormone" (p. 5), and no worse a conceptual "straightjacket" than "organism" (p. 11)!

Regarding therapy, Hesslow is right that disease status is unnecessary as a ground for medical treatment. In the social context, he also rightly denies that all and only diseases should be covered by insurance, exempt people from work or domestic duties, excuse crime, and so on. Even given a basic value framework, a theoretical disease judgment grounds no practical judgments, since the latter depend on details about a disease's severity and effect on a person's unique situation. I do expect in many cases that practical health/disease concepts will arise by our adding criteria to the basic judgment of theoretical disease. A simple example is the Anglo-American insanity defense, in which all four insanity tests begin with the notion of "mental disease" and add further conditions. Of course, one can argue (Fingarette and Hasse, 1979) that legal responsibility would be clearer if we jettisoned mental disease, unifying insanity with other defenses like infancy and intoxication. But there may remain many contexts where we have good reason to aim at a practical version of, specifically, health, and thus to want what I will call a "disease-plus" concept: disease plus extra criteria of severity or disvalue.

To Hesslow's credit, he does not do what so many others do, i.e., take his points either to refute the BST or to show the need for a new, non-disease-based concept of health. Instead he forthrightly calls for abandoning the concepts of health and disease. To reply to such a large thesis needs another paper. However, I agree with much of Nordenfelt's reply (1993) to Hesslow, especially the point that clear analysis is invaluable given the past and present role of the medical vocabulary. Since confusion and abuse of this vocabulary have sometimes been rife, especially in psychiatry, and no analytic consensus is yet in sight, there is more than enough reason to debate the BST and its competitors.

SCIENCE OR BIOLOGY VALUE-LADEN

Finally, many normativist critics deny that medical theory is value-free on the grounds that either science in general or biology in particular is value-laden. My argument was that health is value-free because health is normal functioning, with the normality statistical and the functions biological. This argument admits five lines of attack: to hold that

1. Health is not normal functioning;
2. The normality is not statistical;
3. The functions are not biological;
4. Statistical normality is value-laden; or
5. Biological function is value-laden.

Some writers claim statistical normality in medicine is value-laden, because acceptable performance levels vary with an individual patient and his situation. However, this point confuses theoretical with clinical normality. Other critics argue for the value-ladenness of health by arguing that biological function is a value-laden concept.

To establish the latter—the value-ladenness of biological function—some writers, first, appeal to a general value-ladenness of science. Thus, Agich says the BST rests on "an unacceptably simplistic view of science as value-free." Like medicine, "science, too, is a practice laden with particular value as well as conceptual commitments" (Agich, 1983, p. 39). Actually, I score this criticism as a win for the BST. If health and disease are only as value-laden as astrophysics and inorganic chemistry, I am content. I admit having no sympathy for the view that scientific concepts or knowledge is evaluative. Obviously, we do science, as we do everything, for evaluative reasons. But I do not see why our motives for information-gathering must infect the information gathered, injecting values into science, mathematics, and the Bell telephone directory. However, I leave defending the value-freedom of physics to physicists and philosophers thereof. If the BST shows that health in medicine is as objective as physics, it achieves everything I ever dreamt of for it.

A more common view is that biology is a specially value-laden science, either in general or in certain concepts including "function." Surprisingly many authors think that biology rests on an ultimate commitment to the value of life, or at least its preferability to death. Agich quotes Toulmin to this effect:

The fundamental values of clinical medicine are rooted directly in the basic facts about vital organization, while the basic facts of physiology can themselves be stated only in terms that take for granted the value of the chief vital functions of the organism. (Toulmin, 1975, p. 62)

How could this possibly be true? Biologists study the physiology of all sorts of organisms that hardly anyone wants alive: hyenas, rats, tapeworms, hornets, fleas, ear mites, not to mention pathogenic bacteria. There is no reason to think that an insect physiologist must be committed even to the value of insect life in general, let alone his particular species. On the contrary, he might be funded by Orkin, studying how best to exterminate bugs. One could study human physiology from the same viewpoint, if one were designing chemical or biological weapons. Plainly "the basic facts of physiology" would stay the same.

The idea that biological concepts are inherently value-laden seems to have one of two sources. First, it may rest on confusing our motive for studying something with the thing itself. Though we mostly study physiology in order to discover how the body works so that we can stay alive and well, that is equally true of, say, geology or atmospheric science. We study almost everything for our own benefit. Thus, our major reason for caring what creates and maintains planetary atmospheres is that we will die with the wrong one. But that fact has no tendency to show that planetology or metereology is a science of value-laden propositions. The second source of confusion is an assumption that "life" and "death" are normative concepts. I know no reason to hold that what distinguishes life from nonlife is that life is better. What could it mean to say that it is better to be a bacterium than a mineral, or that bacteria are more valuable than minerals, or anything of the sort? People who find themselves thinking in this way need to revisit Sommerhoff's beast and its dung. The difference between life and nonlife is a factual one, not a matter of our preference for one over the other. In fact, we prefer that many indisputably living organisms die.

But whether or not science or biology in general is normative, in the function literature there is a persistent minority position holding function statements to be so. Even Schaffner (1993) takes this view in a recent chapter reviewing some of the history of the function problem. Clearly, by the BST, insofar as biological function statements are normative, health is normative too. Once again I leave this issue to nonmedical philosophy of science. I merely note that for anyone who seeks to exclude values from scientific knowledge, holding biological function statements normative is a case of the high cost of normativism about health. Another cost that several writers are willing, even eager, to bear is to hold specific disease diagnoses value-laden. Thus, Margolis says that "Peter is tubercular" is a value judgment (Margolis, 1976, p. 240), Engelhardt that "George is a schizophrenic" is (Engelhardt, 1986, p. 174).[40] We earlier saw Fulford making "kidney" a normative term. My conclusion from these views is that given time, an honest and intelligent normativist will find the rope to hang himself. Anyone to whom these views seem implausible can take them as some evidence against normativism about health, insofar as leading normativists find them equally attractive.

Health ≠ Absence of Disease

John Ladd's essay (which cites HTC as one opponent) attacks one of the BST's basic features, that health and disease are "interdefinable" because health is the absence of disease. Ladd calls this "an almost universal assumption" (Ladd, 1982, p. 25) by health-care professionals, philosophers, and the public. By contrast, he argues not only for "positive health" beyond the absence of disease, but also for a more radical position that health and disease are "incommensurable."

Toward the incommensurability of health and disease, Ladd offers the following points. The Greeks had separate gods for health and disease; and on the linguistic side, many contrasts appear. "Health" and "disease" are etymologically unrelated.[41] By contrast with "disease," "health" cannot be pluralized, does not admit different types, embraces poor as well as good health, and is predicated of the whole person. Furthermore, health is often a general and abiding condition, and admits of degrees. Most important, health is dispositional, including being predisposed not to get diseases and to cast them off quickly. Partly as a result, "influences on health" (genetic endowment, environment, lifestyle) are wider than what are usually viewed as "causes of disease" by a medical science used to the "fairly simple kind of causal analysis" in infectious disease. Indeed, Ladd concludes,

> My point is a simple one: *healthy people are less likely than unhealthy ones to get sick.* And the causes of good and ill health are largely social rather than medical. (p. 36)

Ladd sees his idea of health as positive and as "contextualistic" in that, as Dubos and other adaptationists hold, what is healthy (e.g., sickle trait) varies with person, situation, and circumstances (pp. 32, 33).

With apologies for this rapid summary, I can only offer equally quick replies. Etymology and theology are weak reeds in conceptual analysis, especially in science. If "cold" and "heat" were etymologically unrelated and the Greeks had separate gods for each, that would not show that cold is not the absence of heat.[42] "Health" can indeed be predicated, like "disease," of body parts, since we speak of a healthy heart or healthy skin. I do not agree that a blind person or amputee can be "completely healthy." It is true that we speak of bad and good health, but not of bad and good disease. However, that is because health is sometimes generic and other times means only good health, the absence of unhealthy conditions, which can be diseases. "Luck" and quantitative concepts like "weight," "size," and "age" are similarly ambiguous between a spectrum and one of its poles.[43] "Disease" can be a general category (vdS/T, p. 104) with specific diseases the taxa. Diseases can certainly be abiding. As for comparisons, I do not see why "one cannot have more or less of a disease" (Ladd, 1982,

p. 27), e.g., more or less cancer or hypertension, and one can surely have more diseases or fewer. On causal analysis, it is a clear scientific error to ignore all causal factors in infectious disease except the "etiologic agent." In fact, that well-established term seems to me wrong: often what a bacterium or virus actually is is only the pathogenetic agent. The true etiology of infectious disease often involves environmental or lifestyle factors, mediated by stress and other psychological events. But if some physicians have simplistic views of disease causation, that is no argument for positive health.

Ladd's most challenging idea is his loyalty to a positive concept of health as what sounds like a vigor or vitality that confers resistance to disease. As I noted in HTC (p. 553), given the BST's definition of intrinsic health as the absence of disease, one can define instrumental health as what tends to prevent disease. Inspired by physical fitness, one can also define a kind of positive health as superior physiological function or functional capacity, at least until one runs into the problem of incompatible excellences (HTC, pp. 553, 554, 568–570). Since Ladd's notion of health as vigor seems to fall under both headings, to a limited degree the BST can accommodate his view. However, he cites HTC, so I apparently left him unsatisfied. Whether there is such a phenomenon as what Ladd calls positive health, either more than my two categories or something quite different from them, is perhaps partly an empirical question.

Mental Health

Few critics comment on the BST's analysis of mental health, which was its major goal. Farrell (1979) has no criticism of the BST. Rather, he holds exactly DI's view that "mental illness" is a mixed normative-descriptive concept of which the descriptive part is biological dysfunction. Champlin (1981) cites DI as one of eleven discussions with which he is dissatisfied, but he offers no specific criticism. His paper is rich in interesting cases, theses, and historical comments. However, it seems to me to rely too much on ordinary-language judgments and distinctions (e.g., between being ill and being in ill health, p. 476) which, though exquisite, are dubious in themselves and of meager relevance to scientific psychopathology.

Substantive criticisms of the BST on mental health are made by Mischel (1977) and Lavin (1985). Mischel's paper, which closely follows the outline of MH with deeper discussion of specific psychological theories, was cut short by his death. So I cannot tell whether the objections that the paper poses to the BST were expository devices or points of disagreement. For example, Mischel at first suggests that since psychopathology is judged by standards of "maturity," which is a "social and evaluative" concept, mental health cannot be "descriptive" and "value-free" (p. 208). But from his next page and the editor's note (p. 217), it seems that he may have planned to conclude that maturity can, in fact, be judged by an objective "theory of psychological development" (p. 209), as I would say. His other objection (p. 207) is that no species-typical mode of perception and cognition can be identified, since individuals and cultures differ in their cognitive and perceptual styles. Lavin, too, questions whether we can give a "value neutral account of species typical *rationality*" (p. 542). It is hard for me to understand this objection. What is cognitive psychology about, if not species-typical patterns of cognition? I should think rationality is, on the contrary, one of the areas of the human design most accessible to psychological research, especially since normative theories of rationality (logic, probability, statistics) already exist for comparison. If cultural or individual differences emerge, I would expect to treat them as normal variability or, at worst, polymorphism.

There remains Lavin's other charge, that the psychoanalyst's concept of function differs from the physiologist's because "[t]he Id, Ego, and Superego *never fail* to perform the functions assigned them" (p. 542). I am mystified by this thesis, since psychoanalytic psychopathology claims to be based on dysfunction in these three agencies (Brenner, 1974; Holzman, 1970). Psychopaths, for example, essentially lack a superego. Psychotics have an ego lesion that causes a "break with reality" leading to massive irrationality, though Lavin seems to doubt that there are any objective standards of rationality. His discussion prior to this thesis is useful and convincing, but he needs to develop his criticism of the BST at greater length.

A mental-health counterexample several critics pose to the BST is unusually risky behaviour, such as race-car driving[44] or "sensation-seeking" generally (Pawelzik, 1990, p. 19). I am puzzled by these writers' confidence that such behavior is not pathological, since that view is hardly a psychiatric consensus. On the contrary, with due regard for normal variability, psychoanalysis is willing to view some unusually risky behavior as a pathological expression of, e.g., unconscious masochism and guilt. The fact that the career brings honor, fame, and fortune does not block this judgment (cf. CH, p. 379). But by the BST, unlike Kendell's and Scadding's views, a psychopathology judgment should not rest directly on increased risk of death, but on an identifiable mental part-dysfunction. As when squirrels cross the street for food, behavior can reduce survival chances without involving any species-atypical level of psychological part-function.

Failure to Fit Medical Disease Classification

Before we begin specific issues and cases, a word on methodology. I assume that, as Ladd says, "the concept of disease is primarily a technical medical concept" (Ladd, 1982, p. 33); its home is scientific medicine. Thus, *contra* writers like Fulford and Reznek,[45] I see no interesting or authoritative ordinary usage of "disease" that we need consult in analyzing it. A corollary is that hypothetical cases have less force on this topic than is usual in philosophy. Normally, hypothetical cases play a huge role in philosophical analysis. But if we seek to analyze a technical concept of medicine, we should not appeal to what "we" would call a disease, or pathological, unless "we" can predict hypothetical medical usage. At a minimum, nonphysicians should be wary of this task. There is scarcely more reason to pay attention to what philosophers or laymen would call a disease than, say, to what they would call a tort. Although I love to use the word "tort," I would not try to define it by my own intuitions about hypothetical torts and nontorts, since those intuitions deserve no respect.[46] Thus, for me Pawelzik (1990) is an ideal critic in confining himself to real cases from the *Cecil Textbook of Medicine,* while Reznek, all of whose main cases against the BST are hypothetical, is at the other extreme. Although I think Reznek's and other writers' hypotheticals give the BST no difficulty, it is still true that real cases are superior in that with them, we can at least be sure what physicians would say.[47]

Dysfunction vs Diversity of Disease Classification

Kendell remarked that current nosology includes disease entities deposited, like shells on a beach, by successive waves of medical science.

> Each of these waves of technology has added new diseases, and from each stage some have survived. A few, like senile pruritus and proctalgia fugax, are still individual symptoms. Others, like migraine and most psychiatric diseases, are clinical syndromes—Sydenham's constellation of symptoms. Mitral stenosis and hydronephrosis are based on morbid anatomy, and tumours of all kinds on histopathology. Tuberculosis and syphilis are based on bacteriology and the concept of the etiological agent, porphyria on biochemistry, myasthenia gravis on physiological dysfunction, Down's syndrome on chromosomal architecture, and so on. (Kendell, 1975, p. 307)

Recently Scadding suggests that this point shows a defect in the BST.

> ... Boorse's definition gives undue prominence to functional deficit as a defining characteristic of diseases, and thus conflicts with current nosology, in which diseases are

defined by several sorts of characteristics, among which aeti-
ology takes precedence. For this reason, I prefer the wider criterion of biological disadvantage. ... (Scadding, 1988, p. 123)

I answer as follows. First, the BST is aimed at the demarcation problem, not at nosology. It seeks to say what is disease, not
to individuate diseases. Individual disease entities should be
defined and classified on whatever is the most scientifically convenient basis. I suppose I agree that etiologic classification is the
best basis where available. However, terms representing pathophysiologic entities, like "diabetes mellitus," will remain useful
even after we discover multiple causes of the condition.[48] At any
rate, if one merely divides disease D into two forms D1 and D2,
by etiology or in any other way, that does not affect health as the
absence of disease, since the absence of D is identical to the
absence of both D1 and D2. Thus, the BST *per se* has no implications for nosology except that any genuinely pathological state
must include dysfunction. It does imply that an individual disease
entity defined on a nondysfunction basis must produce dysfunction in every case, or not (always) be a genuine disease. It is not
clear to me whether Scadding and Kendell disagree with this
claim. Can there be a case of Down's syndrome, mitral stenosis,
Laennec's cirrhosis, syphilis, tuberculosis, hydronephrosis, por-
phyria, myasthenia gravis, and so on, that does not include dysfunction, even cellular dysfunction, yet is surely disease? If so,
that is a fatal objection to the BST.
    Confronted with such a case, I would try to save the BST by
questioning either whether it is a case of the disease named, or
whether that "disease" is a disease in every case. For example, if
trisomy 21 sometimes produced no dysfunction of any kind, then
I would ask whether trisomy 21 is, in fact, always Down's syndrome, or whether Down's syndrome is, in fact, always pathological. On the contrary, I should think that in such people trisomy
21 would be a normal variant, not a disease. Similarly with, e.g.,
minuscule narrowing of the mitral valve. The BST can say either

that it is too minor to be called mitral stenosis, or that a tiny amount of mitral stenosis can be normal. But if a *forme fruste* of a disease is so *fruste* that there is no dysfunction at all, even at the cellular level, then I think it ceases to be pathological and becomes a normal variant which in other people causes disease.[49] I doubt that Kendell or Scadding would disagree. They seem to be speaking of ordinary clinical forms of the diseases they mention, not exotic theoretical variants. Thus, I cherish the hope that Kendell and Scadding can accept the BST as solving the demarcation problem, once that is distinguished from nosology.

Past vs Present Functions

A trait can become fixed in a species by a beneficial effect on survival or reproduction which then stops being beneficial, e.g., because of environmental change. Several authors take this fact as a problem for the BST but, interestingly, make opposite criticisms.

Engelhardt seems to believe that the BST misses the fact that the species-typical, formerly functional trait is now a disease because some species-atypical trait is now a better adaptation:

> One cannot simply turn to the results of evolution to determine what a disease is. We are the product of blind, selective forces, which, if they have been successful, have adapted us to environments in which we may no longer live. Since what is species typical may represent an adaptation to environments in which we no longer live, it may not afford us the same degree of adaptation as that provided by some species-atypical trait.[50]

But Wakefield, classing me in the "biological disadvantage" school, makes exactly the opposite criticism:

> Scadding ..., Kendell ..., and Boorse ... were right that there must be an evolutionary foundation to our judgments of disorder. ... However, the biological disadvantage approach mistakenly uses decreased longevity and fertility in the present environment as the criterion for mechanism

dysfunction. The fact that the organism's mechanisms were originally selected because they increased longevity and fertility in a past environment does not imply that some mechanism is malfunctioning when longevity and fertility decrease in the present environment. Thus, despite its evolutionary roots, the biological disadvantage definition actually fails to require a dysfunction and thus is subject to counterexamples. (Wakefield, 1992a, p. 379)

There are two issues here: defining function and defining disease. Regarding function, a contrast between Wright's view and mine is that I require a function to be an actual contribution to a goal, whereas Wright says that something can "have" a function that it no longer performs. Wakefield, then, is at worst objecting to my view of function. His own view, as I said, is essentially the BST with Wright's function theory replacing mine and a harm clause added to make dysfunction clinically important. On the other hand, I do see a species as extending over time as well as space, so for me some of the past affects what is species-typical. I do not take sudden or temporary changes in lifestyle, even if worldwide, as changes in the nature of the species.[51] For example, if tomorrow all human beings suddenly began to live wholly inside buildings, I would not immediately say that human skin had lost its function of synthesizing vitamin D in sunlight. If the whole earth went pitch black for two days, I would not say eyes had lost all function in the human species (cf. Margolis, 1976, p. 247). While my views on time are vague, my concept of a reasonable time-slice of a species seems analogous to the species concept in paleontology, where (with Mayr's interbreeding test lacking) vague boundaries are the rule.

On most real examples, I expect to side with Wakefield and against Engelhardt. This is no imaginary issue. Human beings today suffer many difficulties rooted in the fact that our physiology, and still more our psychology, evolved to fit a different way of life, such as primitive hunter–gatherer society. Unregulated absorptivity of the small intestine, or excessive love of leisure, causes obesity. Undischargeable aggression causes, perhaps, hypertension and the risk of extinction by nuclear war;[52] and so on. With Wakefield and against Engelhardt, I do not consider these basic traits disease, though they make disease more likely in current environments. Rather, they are part of the human design, as judged over the vast majority of the species' history and even the species' recent history. Contemporary Western civilization is barely an eyeblink in the history of man. So these are further examples of the difference between disease judgments and adaptation judgments.

Internal States

Hare starts out with a narrow concept of disease, noting that all diseases are pathological but not conversely, and that "injuries and wounds" are bad conditions treated by doctors which are not diseases. He then asks: "Why do the attacks of viruses count as diseases, but not the attacks of larger animals or of motor vehicles?" (Hare, p. 175) Hare goes on to mention worms large and small, fleas, lice, mites, and the ichneumon fly, saying that intestinal worms are called a disease but flea or louse infestation is not. He wonders whether "perhaps we use the word 'disease' for conditions whose cause was not visible before the invention of microscopes." Later, discussing me, he asks:

> What does 'internal' mean? As we have seen, a skin disease may be in no stronger sense *in* (ie inside) the organism than are maggots, which are not a disease. They are indeed, a condition *of* the organism; but this wider description will not bear the weight put upon it by Boorse's definition. Being hung in a noose is also a condition *of* the organism; it, likewise, interferes with species-typical contributions to survival and reproduction. The same is true of the condition of being tarred and feathered, and of being bitten by dogs or run over by lorries. None of these is a disease. (p. 176)

Here two issues seem to be run together. One is broad vs narrow uses of "disease." Dog bites and truck injuries are not usually called diseases, but only because narrower uses of "disease" are more common than the broad usage I am trying to analyze. Hare's "wounds and injuries" are certainly pathological conditions, and the BST has no problem with them. Lacerations, fractures, and so on, from dogs or trucks are certainly "internal states" with species-subnormal part-function. Thus, wounds are a red herring regarding internal states.

Other examples are more relevant to that issue. Among Hare's lower organisms, the BST implies that fleas and lice are not a disease state or pathological condition because they are on, not in, the organism. Flea bites are pathological; fleas aren't. They are annoying, like flies, roaches, and teenagers; their bite is a pathological condition; but in themselves they are not a pathological condition because they are not in the organism. Penetration is all.[53] Hare writes:

> Does a disease have to be something *in* me? And in what sense of 'in'? Some skin diseases such as scabies are so called, though the organisms which cause them are in the skin, and do not penetrate the body. They penetrate the *skin* indeed; but then so does the ichneumon maggot, and the body too. Is the difference between these maggots and the scabies mite merely one of size? Or of visibility? (p. 175)

No. Hare goes wrong contrasting "skin" with "body." Skin is part of the body, in fact its largest organ; any other view would be deeply unmedical. So I shall rule on behalf of the BST that all these organisms cause a pathological condition only when they cross the organism's boundary by penetrating living tissue or a body orifice. As far as I can tell, this view fits medical usage well enough.[54]

Hare's remaining counterexample here is hanging. Although hanging is not an internal state, the circulatory and airway blockages caused by the noose pressure certainly are.[55] Since these blockages interfere with physiological functions, indeed are fatal, the BST should count them as pathological conditions. In fact, it seems impossible to cause death without causing a pathological condition. Death is the ultimate pathology, and also, as the BST predicts, the end of all function. The only reason one would not count such circulatory or respiratory blockage as pathological would be to exclude conditions continuously maintained by an external force. That is the difference between hanging and a dog bite. The bite is a wound that outlasts the dog, whereas removing the rope (or the gravity) restores the hanging man, we may suppose, to instant normality. To exclude such conditions from their category of "malady," Culver and Gert (1982, pp. 72–74) required that maladies not have a "distinct sustaining cause" like rope. In CH, I said one might use such a clause in defining pathological condition. It now seems better to admit that some pathological conditions are continuously maintained in an otherwise normal organism by an external cause. Hanging is not in itself a disease or pathological, but it causes a pathological state that is quickly fatal. Except for being deliberately caused by people, hanging does not much differ conceptually from arctic hypothermia, heat exhaustion, or other environmental pathophysiology. I do not see what conceptual problems or clashes with medical usage this ruling produces, as long as one avoids confusing broad with narrow uses of disease.

Hare on Hair

After his pitiless scrutiny of "internal state" and other key concepts of the BST, Hare proposes to "waive these subsidiary difficulties" with my theory, in favor of his main argument about hair (Hare, 1986, p. 178). I appreciate this waiver, since it makes it unnecessary for me to remark how much more kindly he judges his own theory than mine. So I need not observe that Hare not only puts *ceteris paribus* language ("in general") in the normative clause of his account, but then also cites Wittgenstein on family resemblance to excuse the clause, *ceteris paribus* and all, from applying to every case of disease. I can withhold my opinion that family resemblance is the last refuge of a scoundrel, and that referring to this Wittgenstein passage should be made illegal in analytic philosophy.

The hair argument, offered as a query about "natural function," is as follows. Baldness is "naturally" called a disease. However, if genetic engineers developed an organism which, when spread on women's legs, had the loss of leg hair as its sole effect, we would not call this condition a disease, nor would doctors worry about this microbiologic depilatory. Since growing hair is equally a "natural function" of head and leg skin, the contrast in the two cases must be normative. In calling baldness a disease we must partly mean that it is bad for you. "It is the fact that the ladies want to get rid of their hair, but balding men want to keep theirs, that makes the difference." Thus, disease and health are evaluative concepts.[56]

To reply, I am, first, uncertain of the disease status of the two hairless conditions. What pathologists would say about Hare's depilatory, if it had no other effect than leg-hair loss, I do not know. If it caused leg hair to fall off by damaging skin cells, it would seem to be a kind of convenient pathology, much as oviduct blockage is for some women. I also have doubts about the disease status of baldness. To be sure, loss of scalp hair (alopecia) is divided by the AMA *Nomenclature* (1961, pp. 131–142 *passim), the International Classification* (1992, pp. 616,617), and textbooks of dermatology[57] into a surprising variety of disorders. Some types of alopecia are signs of other diseases, such as pituitary or thyroid conditions or Stein-Leventhal syndrome, or of poisoning, radiation injury, and so on. Ordinary male-pattern baldness, which ICD-10 includes under the title of androgenic alopecia, afflicts almost all men to some degree. But that would not rule out an extremal disease of excessive or premature androgenic baldness. Nonetheless, I would wish to be sure of pathologists' determination to call any sort of baldness pathological before I rested the normativity of health and disease on this example. Perhaps alopecia is one of the structural diseases, like macacus ear and aberrant hymens, that I have already said should not be in nosologies at all.

On the other hand, head hair but not leg hair may have a function, in which case the BST predicts Hare's two disease verdicts. Clearly body hair is vestigial in human beings, while scalp and face hair are not. Since, in our evolution from earlier primates, hair has almost vanished from the rest of the body but not the head, that is some proof that it retained a function there. There is no generally accepted view of what that function is or was. One possibility is that head hair has a protective or insulating function. It is often said that in cold weather, 20% of body heat is lost from the head; at the same time, hair also has an effect on heat absorption in hot climates. (On thermal effects of hair, *see* Newman, 1970.) As an additional function, Guthrie proposes that both face and scalp hair in male primates, including man, are threat-display organs designed to intimidate other males by increasing the apparent size of the face and head (1970, pp. 262, 273). However, Guthrie also endorses Goodhart's proposal (1960) that balding is functional, in increasing the forehead area for threat by skin-reddening. Many other possibilities exist, such as that the graying of men's hair, by indicating age, serves as a fitness indicator in sexual selection.[58] At any rate, if human scalp hair but not leg hair has had an evolutionary function, then the BST will count baldness as pathological, but not Hare's natural biodepilatory.

Both the above sexual-selection hypothesis and Hare himself open a can of worms I would have liked to avoid: the reproductive effect of appearance. Yet another point about baldness is that many women find bald men unappealing, while few men demand women with hairy legs. This fact too offers the BST a route to Hare's two disease judgments, if only baldness impedes reproduction. Yet one cannot let the BST turn ugliness into a disease, especially not moderate or marginal ugliness, since it isn't one. But baldness, if ugly, is not simple ugliness; it is the absence of a normal body part, a discrete structural abnormality. One can see how a structural deviation from species design, coupled with damage to reproduction, might induce a BST-inspired physician to call a trait pathological. Perhaps, then, this line of thought is one reason major deformities are seen as patho-logical. Structural defects much worse than baldness can be so hideous as to make reproduction almost impossible, though major structural defects (harelip, cleft palate) tend to involve dysfunction as well as deformity. Still, it seems odd to call attracting the opposite sex a biological (let alone a physiological) function of the mouth, face, scalp, fingers, spine, and so on. So perhaps the BST does not, after all, entail that awful structural abnormalities are pathological, despite their antireproductive effect. One should note that this inference could only apply in any case to deformities that block reproduction throughout our spe-cies. The BST cannot make any one time's or culture's standards of beauty into requirements of health. And no such link between appearance and reproduction offers much comfort to normativ-ism anyway, since a fact about what the human race finds intol-erably ugly, though a fact about values, is a fact nonetheless.

Masturbation and Drapetomania

Since BST critics do not stress cultural variability in disease judgments, I will omit the general form of that issue. Regarding primitive cultures, I suggested in CH that when they classify different things as diseases from us, one must ask whether their term means the same as Western medical science's "disease." If it does, then either we or they are wrong about what is biologi-cally normal in the human species. But what critics do often cite is two or three examples of Engelhardt's—masturbation, drapeto-mania, and dysaesthesia aethiopis—which were disease catego-ries in 19th-century Western medicine itself. These examples' influence is as wide as it is, to me, incomprehensible. On their face they look pronaturalist, not pronormativist. One would think them clear cases of the danger of confusing social values with disease. Yet many authors use them against naturalism to show that even in scientific medicine, the concept of disease depends on values. Many take them to establish not only normativism, but also cultural relativism, about disease. Probably countless undergraduate readers of the Beauchamp-Walters anthology have drawn from Engelhardt's essay the conclusion succinctly put by Merskey:

> Our predecessors saw masturbation as a subject for treatment and as evidence of disease that would cause further disease. *Their* views are not ours. *For* them masturbation was a disease. *For* us it is not. We cannot escape such relativism. (Merskey, 1986, p. 223)

The masturbation example is invariably cited to Engelhardt's 1974 essay, which is dense with factual detail. Summarizing and extending an already rich historiography of 19th-century masturbation theories, Engelhardt also drew philosophical morals from the episode. But I find it mystifying how anyone can take this history as evidence against naturalism and for normativism. To see why, let us review Engelhardt's story.

After an anonymous Dutch book *Onania* of 1700, Tissot's 1758 book established the doctrine that masturbation, like excessive sex in general, had a debilitating effect. For Tissot this effect was caused by loss of seminal fluid, but later writers focused on the deleterious effect of "nerveshock" on the "nerve-tone," or similar categories of 19th-century neuromythology. A long list of sequelae was found to result from masturbation, including (to name only a few) epilepsy, blindness, deafness, rickets, spinal-cord damage, idiocy, insanity, and ultimately death. Masturbation also caused a certain physical appearance: female genitals enlarged, male genitals shrank, and the patient developed moist hands, stooped shoulders, acne, and a distinctive gait. Not surprisingly, this syndrome of symptoms and signs, with a known cause and well-defined complications, came to constitute a "disease entity," indeed a "fairly well evolved" one (Engelhardt, 1974, p. 237). Physicians as eminent as Rush, Hutchinson, and Freud accepted the general theory. And why not, since "the categories of over and under excitation suggest cogent, basic categories of medical explanation" (pp. 239, 240)? In turn, a wide variety of therapies was directed at the disease, including such "drastic" or "heroic" remedies as acid burns on the genitals, needles in the testicles, prostate, or bladder, and electrodes in the rectum. Some men were castrated, and Isaac Baker Brown, the most famous surgeon of this kind, cured female masturbators' "losses of nerve force" once and for all by chloroforming the patient, then cutting off her clitoris "either by scissors or by knife—I always prefer the scissors" (p. 244).

The first point about this chronicle is, of course, that 19th-century masturbation science consists of spectacularly false statements. Masturbation does not damage the nerve-tone, much less the spinal cord. It does not cause rickets, blindness, insanity, acne, or even moist palms. Theorists were wrong in their theories, wrong in their data, wrong about virtually everything; almost no statement listed in the essay contains even a grain of truth.

But falsity aside, the reason this history is hard to see as supporting normativism about disease, much less relativism, is simple. As Engelhardt shows, the disease status of masturbation was embedded in and supported by an array of ordinary causal statements about its effects. This means, first, that his story offers no more *prima facie* reason to be a normativist or relativist about disease classification than about causation. Merskey says that "for them" masturbation was a disease. Equally, "for them" masturbation caused acne, blindness, insanity, and death. Such language resembles the locution "true for X" beloved of philosophy students, whose instructors duly point out that by "p is true for X" they mean no more than "X believes that p." Similarly, why should we describe the history otherwise than to say that 19th-century doctors believed that masturbation was "a disease that would cause further disease" and were wrong on both counts? Why is one error factual, the other normative? Normativists assume that disease status is normative, but Engelhardt's story does little to prove it. Second, as he observes, if the elaborate causal theories of the pathology and sequelae of masturbation had, in fact, been true, then masturbation (or the internal state that caused it[59]) would have been a disease by current scientific models, indeed perhaps by the BST itself. So once again it seems that 19th-century physicians simply made two scientific errors: false causal beliefs about masturbation and a false belief about its abnormality.

To compare disease status to causation is no eccentric view of Engelhardt's story. On the contrary, his own interest in the tale is precisely to teach a general lesson about values in science. His focus is not only, or even mainly, on disease classification, but on epistemology in medicine and science generally. In paragraphs early and late, he draws the following general morals:

> [O]ne chooses concepts for certain purposes, depending on values and hopes concerning the world. The disease of masturbation is an eloquent example of the value-laden nature of science in general and of medicine in particular. In explaining the world, one judges what is to be significant or insignificant. ... The problem is even more involved in the case of medicine which judges what the human organism should be (i.e., what counts as "health") and is thus involved in the entire range of human values. ... (p. 234)
>
> In both [centuries], expectations concerning what should be significant structure the appreciation of reality by medicine. The variations are not due to mere fallacies of scientific method, but involve a basic dependence of the logic of scientific discovery and explanation upon prior evaluations of reality. ... The disease of masturbation is an eloquent example of the role of evaluation in explanation and the structure values give to our picture of reality. (pp. 247, 248)

I cannot here discuss normativism as a general view about science, holding that science is permeated with values even in simple causal claims like "masturbation causes rickets." Engelhardt's philosophy of science, which he cites to Hanson, is at best a minority view. Here it suffices to note that his essay's thesis is precisely that disease status is value-laden in the same ways as causation and science in general. And as noted earlier, if "X is a disease" is only as value-laden as the rest of science, e.g., astrophysics, I am content.

Engelhardt does make two points that could show disease statements more normative than the rest of science. First, he suggests that the disease status of masturbation preceded the causal theories, with masturbation becoming a disease when a moral value was translated into a medical judgment. So far, this view fits the BST perfectly. The obvious view for a BST supporter would be that moral condemnation of masturbation came illegitimately to be cloaked in a false medical claim of pathology (Reznek, 1987, p. 128; Daniels, 1981, p. 156; Munson, 1981, p. 201). But Engelhardt too distinguishes moral from medical judgments, as two different sorts of evaluation. He thinks health values are not moral ones, but the esthetic values of "natural teleology."

> This analysis, then, suggests the following conclusion: although vice and virtue are not equivalent to disease and health, they bear a direct relation to these concepts. Insofar as a vice is taken to be a deviation from an ideal of human perfection ..., it can be translated into disease language. ... The shift is from an explicitly ethical language to a language of natural teleology. To be ill is to fail to realize the perfection of an ideal type; to be sick is to be defective rather than to be evil. The concern is no longer with what is naturally, morally good, but what is naturally beautiful. (Engelhardt, 1974, p. 247, 248)

Again the next year Engelhardt repeated that "Health is a normative concept but not in the sense of a moral virtue." Rather, "[h]ealth is more an aesthetic than an ethical term; it is more beauty than virtue."[60] To this I reply that in Sommerhoff's sense, "natural teleology" is not esthetic judgment, but scientific fact. What makes beast beast and dung dung is not that one is prettier than the other. It is that one is objectively goal-directed and the other is not.

In the end, to argue for the normativity of disease, or for much of anything, by appeal to cognitive atrocity stories from what are usually seen as the 19th century's two worst scientific disasters, its sexual and racial science, is an odd strategy. Engelhardt offers little reason to see the theories he describes as anything but bad science supporting awful therapy. Except for masturbation mania's wide acceptance, Isaac Baker Brown is comparable to William Arbuthnot Lane, another Victorian surgeon-theorist, who spread similar human misery by attributing all manner of pathology to irregular bowels, then solving the problem by colectomy.[61] Usually, one traces the beginnings of a genuine science of sex to pioneers like Krafft-Ebing, Ellis, and Freud in the closing decades of the century. Moreover, I think the evidence will show that what finally dethroned masturbation as disease and cause of disease was not a change in its moral status—after all, the fount of moral condemnation, the Catholic church, still views masturbation as sin—but simply the first reliable data on its prevalence.[62] After Kinsey (1948) proved the near-universality of adolescent masturbation (that, as the joke goes, statistics show that 99% of teenage boys masturbate and 1% are liars), two things became obvious. First, the old catalogue of sequelae was wholly mythical. Second, if adolescent masturbation is nearly universal, it looks like a normal stage of development. So masturbation is actually a good case for the importance of statistics in medical normality and the unimportance of values. Statistics, not social value-change, ultimately refuted 19th-century masturbation science.

In sum, Engelhardt's essay shows that if we abstract, as he does, from all questions of truth and falsity, then masturbation and drapetomania were Victorian diseases. Equally, if we abstract from all questions of truth and falsity, then cows jump over the moon. Or, more charitably, the example shows that if one is wrong about all the other facts, one can be wrong about what is a disease. But that is trivial. That such an example impresses so many writers on health seems a sad commentary on philosophy of medicine.

Another puzzling aspect of this example is that some writers fault the BST for its unconcern with real patients' "suffering"

(Engelhardt, 1986, 169 ff.) or "pain" (Wulff et al., 1986, pp. 53, 54), yet some of the same authors cite masturbation in support of their allegedly more humanistic normativist account of disease. Do such writers suppose that it was not real men and boys whose penises were burned with acid, and not real women and girls whose clitorides Dr. Brown's scissors snipped off? When such treatments were applied to minors and other defenseless patients, the story is one of medicine using abysmal science to excuse serious human-rights violations. The medical torment and surgical mutilation of masturbating children is one of the vilest chapters in the history of Western medicine. Perhaps normativist writers who cite masturbation are confident we will avoid similar abuses today because our values are so superior to Victorian ones. I have suggested that it was scientific, not moral, change that led to masturbation's redescription as normal. Given a choice as to which sort of progress to rely on, scientific progress is a much better bet than moral. Science, or empirical rationality in general, is not inconsistent with true humanism, but a precondition for it. If the masturbation example shows anything, it shows the BST's value as a bulwark against just such eruptions of malignant moralistic claptrap in medicine. At least, if disease language had been unavailable against masturbators, doctors might have had to ask themselves what right they had to mutilate helpless people to stamp out sexual sin.

Dynamic Equilibrium and Environmental Disease

We come now to a group of four more challenging objections, the first two of which concern functional normality under environmental change. Obviously, no fact is more pervasive than what is often called the "dynamic equilibrium" of normal physiology: the normal functional variation within organisms acting and reacting to their environment. The normal level of almost all part-functions varies with what an organism is doing, what other part-functions are being performed, and the environment. Heart rate, blood pressure, respiration, and countless other variables vary from exertion to rest. The secretion of digestive enzymes is coordinated with meals, sweating with body temperature, and so on. Many functions are performed only intermittently; when performed, they raise or depress other functions. Indeed, a bimodal functional pattern is anatomically enshrined in the division between sympathetic and parasympathetic nerves, a contrast in turn endocrinologically duplicated, complicated, and regulated. Though I did not stress the dynamism of normal physiology in presenting the BST, I always assumed it and agree with Nordenfelt's (1987, pp. 25–30) and Reznek's (1987, pp. 129, 130) remarks on the point. Some critics see in dynamic equilibrium a phenomenon that must derail the BST's value-free definition of normality. Since the health sciences constantly study the limits of physiologic variation as a scientific question, the topic seems unlikely to cause the BST to collapse. I will discuss two forms of the problem, adaptation to the physical environment (vdS/T) and defense against injury (Nordenfelt).

Organisms, including man, have evolved many adaptive mechanisms to maintain normal physiology against environmental variation. As a rule, one finds short-term and long-term mechanisms, or fast and slow ones. A common pattern is that environmental stress evokes short-term compensatory functions that maintain homeostasis up to a point, but beyond that point the coping mechanisms break down and a discontinuity, a discrete state of illness, results. On the other hand, a more gradual environmental change would have given long-term functions time to operate, with the result that a stress level causing illness in the unadapted organism causes none in an adapted one. In such a situation I would wish the BST, like the health sciences, to take the following position. The short-term adaptations are normal variation up to the point of breakdown; the homeostatic breakdown is pathological. As for the long-term functions, both the unadapted and the adapted states are normal, the former being simply a normal vulnerability to disease; but an incapacity to make long-term adaptations is pathological.

Some of this pattern is shown by Engelhardt's Norwegian in Africa. White skin in Africa is not itself a disease. But if the Norwegian is transported suddenly and left in the equatorial sun, he will get at least a severe sunburn, which is pathological. However, if his level of sun exposure had changed gradually, his skin would have tanned, allowing him to stand much more sun without injury. To use HTC's terminology (p. 553), both his original white skin and his tan are intrinsically healthy; the tan is also instrumentally healthy in protecting him from injury. That is, neither tan nor lack of tan is disease, but the latter makes disease more likely in a certain environment.

To confirm the pattern, let us look at vdS/T's two examples and medicine's view of them: heat and atmospheric pressure. The human body narrowly regulates its internal temperature; its physiology can tolerate only about a 3% range of variation in total body heat (Sodeman and Sodeman, 1974, p. 928). A heat load, caused by a hot environment or exertion, stimulates the thermoregulatory center of the hypothalamus to trigger short-term heat-loss functions. Peripheral capillaries dilate, with a rise in cardiac output, to transport heat from core to exterior for radiation or convection. If this is inadequate, sweating occurs, to shed heat by evaporation. These are normal functions, occurring on their statistically normal occasion, a rise in core temperature. Above a certain heat load, homeostasis fails and the person's condition is called pathological. Textbooks recognize a spectrum of heat-related illness, ranging from a "transient, benign disorder" (Wyngaarden and Smith, 1985, p. 2306) such as heat edema or heat syncope, to the more serious heat exhaustion, to the potentially lethal heatstroke. Heat exhaustion involves such symptoms as fatigue, weakness, giddiness, anxiety, nausea, and impaired judgment, and such signs as dehydration, vomiting, hyperventilation, tetany, and a rise in body temperature of several degrees. As this description suggests, the transition from normal stress to heat exhaustion or heatstroke is a clear discontinuity, with not just temperature regulation, but also many other functions (e.g., muscle control), suddenly falling far below species-typical limits.

There are also long-term adaptations to heat involving the cardiovascular, endocrine, and exocrine systems. After 1–2 weeks, a man who on first exposure to a hot environment could not work hard or long becomes acclimatized. His sweat volume and electrolyte concentration and his heart rate at a given work load have all decreased, his work output is nearly what it was in cooler days, and he is much less likely to suffer heat-related illness. This does not, however, mean that the unacclimatized man's state was pathological, but merely that acclimatization makes pathology less likely in this environment. An inability to acclimatize would, however, be pathological, since it would be a defect in the long-term adaptation functions.

We find the same pattern in vdS/T's other example, adaptation to altitude. Persons suddenly transported to high altitudes may suffer a hypoxic disorder called acute mountain sickness. This disorder can include headache, exertional dyspnea, malaise, anorexia, nausea, vomiting, diarrhea, abdominal pain, insomnia, impaired judgment, cyanosis, Cheyne-Stokes breathing, and tachycardia. The patient can also die of acute pulmonary edema. The nonfatal manifestations usually go away by themselves in a few days, but can be treated by oxygen therapy or descent to lower altitude (ibid., p. 2288). Most of these symptoms and signs clearly constitute or imply species-subnormal function. A more gradual ascent, however, allows an acclimatization process to produce more red blood cells. Afterward, a person can live comfortably at much higher altitudes, though at some cost in risk of other pathology. Sodeman and Sodeman comment:

> The erythrocytosis experienced by high altitude dwellers is probably the most common of the secondary polycythemias and it must be considered an appropriate physiologic adaptation rather than a pathologic disorder. However, sustained physiologic adaptations are usually achieved at a certain biological cost, and individuals at high altitudes pay for an

enhanced oxygen transport by problems related to hypervolemia, hyperviscosity, and hyperventilation. (Sodeman and Sodeman, 1974, p. 546)

The authors also remark that "compensatory reserves are ... decreased and the effect of cardiopulmonary disorders must be more serious than at sea level." Finally, above about 14,000 feet, one risks a condition called chronic mountain sickness or Monge's disease, "characterized by what appears to be an exaggerated adaptive response to altitude." Its manifestations include severe polycythemia, low arterial oxygen and high carbon dioxide, and impaired sensitivity to hypoxia, as well as "cyanosis, dyspnea, cough, palpitations, headache, giddiness, muscular weakness, pain in the extremities, sensory and motor changes, and episodic stupor" (Wyngaarden and Smith, 1985, p. 2288). Thus, medicine recognizes an acute sickness caused by sudden altitude stress; a chronic condition of normal adaptation which raises the risk of pathology; and a chronic disease.

Since mountain sickness, heat exhaustion, caisson disease, and so on are typical medical examples of environmental disease, one would expect any conceptual problems in the BST to surface here. I am not entirely sure what problems critics see, but I will try to state and answer two objections. One criticism might be that environmental effects make the normal–pathological boundary vague. This problem does not surface in our two examples. In each case where medicine recognizes an environmental disease, a fairly clear discontinuity has occurred, involving abrupt failure of homeostasis. Sweating and a rise in heart rate are obvious compensatory functions triggered to maintain central homeostasis. By contrast, the symptoms of heat exhaustion or heatstroke are not functions at all, but the breakdown of function. However, even if environmental variation created a vague normal–pathological boundary—suppose that, like some poikilotherms, our whole physiology just ran slower and slower as temperature dropped—I would simply invoke the usual principle and call a tail of the distribution pathological. Vagueness is no more threatening in an environmental case than for any other cause of dysfunction.

Secondly, however, the critic's plan may be to impale the BST on the horns of the following dilemma. On the one hand, if the BST uses pure statistical species-typicality, ignoring where the organism is and what it is doing, it will incorrectly call a heart rate of 150 pathological even during exercise. VdS/T seem to assume that this is my view, though I never dreamed of holding it. On the other hand, if the BST admits that a 150/min heart rate, though statistically atypical, is normal on the occasion of exercise, or sweating on the occasion of hot weather, as I always supposed, then every reaction of the organism will be normal on the occasion of whatever provokes it, and all disease manifestations will become normal—exploding the BST, which is left holding Sober's position that everything possible is equally natural. Now the problem is not vagueness. It is that heat-related illness and even death are precisely the organism's statistically normal responses to an environment of, say, 140° or 200°F, respectively.

The most this line of argument could establish is that Daniels was right (1981, p. 156n) that the BST must introduce a concept of a normal environment. I originally thought of doing so, but decided that averaging over a whole species would exclude deviant environments automatically. I also had in mind the contrast drawn above between short-term adaptive functions and homeostatic breakdown. If, as I now incline to think, these two points do not suffice to handle environmental disease, then the BST needs a concept of normal environment. But it can easily define one in its usual value-free statistical way. Clearly, there are statistically abnormal environments for any species. Equally clearly, medicine maintains that it is impossible to live normally in some of them. Exertion is an inevitable part of the life that is statistically normal for our, or any, species, so there is no prospect of calling that occasion itself abnormal. But living at 14,000-ft altitudes or 140-degree temperatures is not inevitable, but quite atypical for our species. So, at worst, the BST needs to add a rule that a statistically species-subnormal function (in the usual sense of an arbitrarily chosen lower tail) is pathological if it results from an environmental factor outside an arbitrarily chosen central statistical range of that factor in the environments where the species lives. This rule would supplement, not replace, the definitions quoted earlier. Although it would complicate the BST, it does not seem to change its value-freedom in any way.

Defense Mechanisms

Nordenfelt rests his rejection of the BST on one environmental-stress case: defense mechanisms. After illustrating dynamic equilibrium by cold weather and jogging, he turns to "the paradigm case of an infection" to argue that the BST cannot divide normality from pathology.

> A number of pathogenic agents enter, say, the throat. They start producing toxins, which immediately destroy a great number of cells on the mucous membranes. The body quickly reacts to this attack. There is a great concentration of blood at the focal points of the infection; the body temperature rises; certain tissues create antibodies against the viruses and the pathogenic toxin. As a result the toxin is gradually neutralised and the microbes killed. (Nordenfelt, 1987, p. 30)

The problem is that these phenomena (i.e., inflammation and the immune response) are just another "species-typical reaction to a serious attack on the body."

> It may very well be that every part of the body has, during this process, given its species-typical contribution to the ultimate goals of survival and reproduction. In fact, the infectious disease can be seen as the species-typical reaction to the circumstance of a certain microbial invasion.
> But this, then, becomes paradoxical. A typical disease can be seen, on the BST, as a species-typical reaction, i.e.

as a healthy response to a difficult environment. (Nordenfelt,
1987, p. 30)

I was surprised to see Nordenfelt give this defense-mecha-
nisms objection so much weight, since it is easy to answer compared to other environmental cases. Inflammation and the immune
response are unquestionably normal functions, the two most general of the body's defense mechanisms against injury or invasion.
That they are normal is a medically important point, immortal-
ized in Sydenham's famous observation that the manifestations
of disease are the resultant of attacking and defending forces.
This point shows, for example, the clinical danger of treatment by
symptom suppression. Accordingly, the inability to mount an
inflammatory or an immune response (as in cortisone treatment
or AIDS, respectively) is pathological, precisely because it is the
failure of a normal function.

The answer to Nordenfelt is twofold. First, the defense
mechanism *per se* is medically normal, not pathological, just as
the BST implies. Thus, an infection would be more pathological
without it than with it. Second, what is pathological in the infection is the invasion and injury, to which the defense mechanism
is a normal response. Nordenfelt notes that the virus and its toxin
"destroy a great number of cells on the mucous membranes."
That is a straightforward injury, the statistically abnormal death
of a body part, hence is pathological by the test of cellular dysfunction. It is not true that "every part of the body" has performed
its species-typical function; "a great number" of them have died.
Nordenfelt earlier (pp. 27, 28) discounted this point in rejecting
the "interpretation" of the BST that carries dysfunction all the
way down to the cellular level. (That is no interpretation, since
HTC was clear on this point.) His reason was that if one dead cell
is disease, then "we would all be ill" (p. 28). Of course we are not
all ill, but we do all contain some pathology, of which one dead
cell is just a trivial example. If microbial attackers have destroyed
a large number of cells, no pathologist would hesitate to call that

pathological. So we have clear dysfunction. Defense mechanisms simply belong to a class of functions triggered by dysfunctions, a class that exists because organisms have evolved some ability of self-maintenance and -repair.[63]

Before leaving environmental stress,[64] I must comment on the BST's clause on universal environmental injuries. I wrote it to help with the anomaly of universal diseases, but it has been much attacked. Hare and others complain that everything in the organism is the joint effect of heredity and environment; vdS/T (1988, pp. 86, 89) spend a long time attacking the BST for drawing any such distinction. I am much more optimistic than they about the usefulness of distinguishing environmental and internal causation (cf. Note 27 on Sober). But it is unclear that the whole issue need arise, since it is unclear that my environmental-injury clause was ever worth the trouble. HTC said that many universal diseases could already be called disease because they were atypical at any specific body location, and, about athero-sclerosis, that it might be more correct to say the disease is pre-mature, or age-excessive, atherosclerosis. That left only tooth decay (which some say is not universal) and lung irritation. Per-haps one should not bifurcate a definition for the sake of only a small part of a class (universal diseases) which is itself an infini-tesimal part of the field (medically recognized disease). If the internal/external distinction is scientifically problematic, the BST can drop the clause and say medicine should not recognize these universal diseases after all. Alternatively, it can retain the clause and say that in this respect the BST caught medicine using out-moded science. Clearly the critical literature, in seizing on this small point, has greatly exaggerated its importance.

Polymorphism, Heterosis, and Adaptation

An important cause of morphologic variation in a species is discrete polymorphism, which HTC illustrated by eye color and blood type and said the BST could handle disjunctively. An important special case of polymorphism is called balanced polymorphism, heterozygote superiority, or heterosis. The most famous example, used by almost every BST critic, is the balance in the human species between normal hemoglobin (HbA) and hemoglobin S (HbS). I will review the facts and then analyze the example.

HbS is so named for its tendency to crystallize, under low oxygen tension, in a rigid cylindrical form that deforms the red blood cell into a sickle shape and leads to its destruction. In persons homozygous for the HbS allele, this sickling occurs massively and often. Such "sickling crises," where many blood vessels are clogged by sickled cells, cause significant pain, injury, disability, and, without medical treatment, death before reproductive age. This condition is the disease sickle-cell anemia. Heterozygous persons (HbA/HbS), who are said to have "sickle trait," have less of the abnormal hemoglobin in their red blood cells. Now another trigger of sickling is infection of an erythrocyte by the malaria parasite *Plasmodium falciparum*. Plasmodium-infected blood cells sickle, attach themselves to the vessel wall, and are eliminated by the body along with their parasite. Thus, in effect, the HbS creates a self-destruct function in infected cells. Consequently, in high-malaria environments, sickle trait offers defense against malarial infection, making the heterozygote fitter than either homozygote and preserving the sickle gene in the population. One study of a Nigerian population found 76% to be HbA/HbA, 24% HbA/HbS, and 0.2% HbS/HbS, and estimated the normal homozygote's fitness at 88% of the heterozygote's and the HbS homozygote's fitness at 13%. Among American blacks, about 10% have the sickle trait, while one in 400 has sickle-cell anemia (Wyngaarden et al., 1992, p. 888). Sickle trait causes no major damage except in low-oxygen situations, when a sickling crisis can occur. Consequently, heterozygotes enjoy much greater resistance to malaria, at the cost of decreased ability to function at a high altitude, e.g., in the mountains or in jet cabins.

Undoubtedly, sickle trait makes one better adapted to a malarial environment, while the normal homozygote is better adapted to malaria-free mountains. But the first point to see is that, as HTC noted, superior adaptation is not *per se* superior health, since undisputed pathology is sometimes advantageous in special environments. One of Engelhardt's examples (1986, p. 166) is the advantage of color-blindness in environments where it is important to spot camouflage. He implies that only by appeal to God's plans could one consider color blindness "truly and objectively" a disease contrary to natural design. But one can easily extend this case to wholly blind people's advantage in pitch-black environments. There are many situations or jobs (in caves or Vietcong tunnels) where the superior nonvisual senses of the blind will be advantageous. Contrary to Engelhardt, blindness remains pathological by the concepts of a wholly secular medicine. After all, every disease D is advantageous in some environment, if only in a Nazi death camp whose medical director is doing private research on curing D. Health is a matter of species-typical adaptations, not adaptedness to special environments. Using adaptation alone, the normal-pathological boundary cannot be drawn at all.

Thus, the health/disease and adaptation/maladaptation contrasts should not be confused. Writers like Engelhardt, Bechtel, Ladd, Reznek, and many others who collapse the two concepts may be confusing theoretical judgments with therapeutic ones. But to use adaptedness even as a therapeutic guide is dangerous, as noted in my remarks on Ryle's Durham miner (HTC, p. 549) and Lady Wootton's on the defects of the "adjustment" ideal of mental health (MH, p. 69). For it may be the environment that needs to be changed.[65] Adaptationist views of health seem to me much like the current trend toward confusing "may" and "might": where our language offers us the wealth of two concepts, these writers insist on throwing it away for the poverty of one.[66]

But if health is not environmentally relative adaptation, what should the BST say about sickle trait (SCT) and sickle-cell anemia (SCA)? SCA is clearly a disease, to medicine and to the BST. The sickling crises, which are species-atypical, cause circulatory microdysfunction which leads to gross dysfunction in affected organs, and life expectancy is lower even in malarial areas. SCA is a disease maintained by heterosis. But that cannot stop it from being a disease unless one confuses two concepts of "species design": species-typical design of individuals vs design of the species as a class. The former is the BST's concept, and under it, the HbS/HbS genotype and the SCA phenotype clearly are contrary to species design, i.e., they are species-atypical and cause dysfunction. The fact that in some sense the species, as a population, may have a design including this balanced polymorphism does not count.[67] What of sickle trait? It too should be a disease by the BST, since it too is species-atypical and is disposed to reduce the efficiency of a species-typical function (circulation) in low-oxygen environments. Indeed, it shows demonstrable pathology: renal papillary microinfarction is "universal" in SCT (Wyngaarden et al., 1992, p. 891), and many people with sickle trait also have painless hematuria or impaired ability to concentrate urine (Wyngaarden et al., 1992; Isselbacher et al., 1994, p. 1736). Hence, sickle trait looks like a disease that protects you from another disease, as cowpox does against smallpox.[68]

Since that is also the view of current medicine, sickle trait is a clear confirmation of the BST. Sickle trait appears in both the *Nomenclature* (Thompson and Hayden, 1961, p. 230) and *ICD-10* (WHO 1992, p. 253). The *Oxford Textbook of Medicine* calls it a "sickling disorder" (Weatherall et al., 1996, p. 3513), and the *Oxford Medical Companion* treats sickle trait under the heading "sickle-cell disease," which it says is "fully expressed only in the homozygous state" (Walton et al., 1994, p. 902).

On the other hand, though the BST here succeeds in matching medicine, one may feel unhappy with calling SCT pathological. Such dissatisfaction derives, perhaps, from a sense that malarial environments are no less common than low-oxygen environments. Why, after all, should a variant that raises one function while lowering another be called disease, any more than superior health? Because such environments are common, some would prefer to call SCT a normal variant advantageous in certain environments. But then it is awkward to recognize a normal variant that exists only on condition of the existence of disease in other persons. Given the range of possible heterozygote examples, I suspect the truth is that in heterosis, the medical concept of disease is beginning to show signs of strain. The more heterosis comes into play, the less clear or useful the normal–pathological distinction is likely to become, and the more we may prefer instead to talk, like Bechtel and Engelhardt, of adaptation. But it will remain a virtue of the BST to explain the limits of the normality concept, showing where it is more useful than others like adaptation, and where and why it is less so. To use the same tool for all problems simply ruins the tool.[69]

Aging, Menopause, and Goal Conflicts

Many writers take phenomena of aging to refute the BST—either actual phenomena like menopause and senile dementia, or hypothetical ones like self-destruct functions. These criticisms stem from two features of the BST. First, the BST relativized normality to age to accommodate normal development, e.g., the fact that normal babies cannot walk or talk even though most humans can. Consequently, I held that "normal aging" cannot be a disease. Second, the BST includes reproduction as an independent goal, which raises the specter of conflict with the other goal of individual survival.

Let me deal with actual phenomena first. Miller Brown worries about the biological role of "individual survival after reproduction"—e.g., that perhaps after menopause no physiological process is "evolutionarily relevant" (Miller Brown, 1985, p. 316). Reznek says the same (Reznek, 1987, p. 131). That may be an issue on Wright's analysis of function, depending on whether one can still say "the heart is there because it pumps blood" in a postmenopausal woman, given that all selection of women's hearts was on premenopausal women. It is no problem at all on my analysis of function, since the heart continues to make a causal contribution to the woman's survival, to which her physiology is still goal-directed. Thus heart disease remains dysfunctional in older women. Hare notes that most 120-year-olds are no longer breathing, so a man who got pneumonia on his 120th birthday would have no disease (Hare, 1986, pp. 177, 178). My answer is that I assumed the reference class to include only live organisms of a given age and sex. A dead organism is not a normal organism, so any sensible generalization to a biological design must use live ones. (Of course, most members of any species are dead and mouldering, and my reference class includes past members, but I only count how their parts functioned while they were alive.)

VdS/T assert that:

> The distinction of natural ageing *versus* age-related disease … is not as neat as Boorse suggests. Research on age-related dysfunctions continually generates distinct diseases where medicine formerly recognized one homogeneous phenomenon of normal ageing…. Many conditions which are now known as diseases, e.g. rheumatoid arthritis, arthrosis deformans, arteriosclerosis and Alzheimer's disease, were at one time all included in normal ageing under the usual gerontological criteria of universality, deleteriousness, progressivity and irreversibility. And Boorse's criteria of age-related normality and internal causation would in the past have led to the same result. (VdS/T, 1988, p. 87)

This is not a fair criticism. My criterion is age-related functional normality of every body part down to the intracellular level. Obviously, late-20th-century physicians know more physiological detail than 19th-century ones. Therefore, they can identify new diseases, seeing abnormality where a grosser view saw none. Alzheimer's disease was not discovered until the 20th century. Since it is not species-typical even at fairly advanced ages, nor is any of the other main types of senile dementia (e.g., vascular, Parkinson's, multiple sclerosis, Huntington's chorea), what previously looked like normal mental decline now looks like an abnormality in every case. That is all to the good. I welcome any opportunity to call senile decline pathological. But there is no reason why the BST, plus factual knowledge at any given scientific moment, must generate a fixed stock of diseases or infallible knowledge of what is normal.

Conceivably, a similar process will occur with postmenopausal problems like osteoporosis. In any case, many writers claim about both aging and osteoporosis that insofar as such conditions are or become treatable, it is fair to call them diseases, as the BST heartlessly refuses to do.[70] This criticism is wrongheaded. Medical treatment of a condition is neither necessary nor sufficient for it to be a disease. Doctors already treat fertility, pregnancy, penile foreskins, ugly noses, small breasts, pain in childbirth, and so on. Prescribing contraceptives does not make fertility pathological, nor does it have to be pathological for doctors legitimately to treat it. Nothing in the BST implies that medical science should not make every effort aggressively to retard or even eliminate aging, and also to find a way to treat postmenopausal osteoporosis or vaginal atrophy without harmful side effects. The BST is not heartless about treatment; rather, critics are confused about disease. There is no need to ruin the disease concept to help aging men and women, and no point in doing so.

We come now to hypothetical examples, especially the self-destructive functions first proposed by Goosens. Goosens began this criticism by stressing the awful sacrifices that reproduction imposes on some species. The female octopus dies of starvation while guarding her eggs; the males of some salmon species cannot eat because of a hooked jaw adapted to compete for females. Reznek adds several other biological examples of parents dying in the reproductive process, such as the male praying mantis, who must lose his head to ejaculate (Reznek, 1987, p. 111). Goosens then argued against me as follows:

> Suppose in an intelligent and reflective species a structure
> evolved which, about twenty years after sexual maturity,

altered the immunological system so that blood cells ...
attacked the membrane of the brain. The individuals, let us
suppose, suffer greatly and eventually die. Surely they die of
a hideous genetic disease, whose only distinction from others
is its timing, latency, and frequency. It makes no difference to
this judgment if we add that the condition was spread by natural selection and that it continues to promote the reproduction
of the individual and even the survival of the species. The
normal functioning of the bodily parts can then be a disease
because of the harm produced to the individual. Likewise, if
the function of a hormone is to regulate the rate of aging, the
disruption of this function, say by infection or drugs, would
not be a disease if its effect were only to extend a robust life
by delaying a normal rapid acceleration of aging. Or suppose
that the reflective species, in order to reproduce, had to
undergo an irreversible metamorphosis into a sluglike creature.... Whether they are produced by functioning or dysfunctioning should not and would not make any difference in the
basic assessment or response of the medical profession to
these afflictions. (Goosens, 1980, pp. 112, 113)

Reznek, too, rests his rejection of the BST mainly on self-
destruct functions and other conflicts between individual survival and reproduction.[71] He claims that at one time goiter in Italy
was considered so attractive that it increased; hence it had a function, but was still a disease. (Of course, the dysfunction that makes
goiter a disease is still present, so this seems to be a case of goal
conflict.) Reznek also cites the proposal of some writers that
female orgasm has no function. Even so, if a viral infection interfered with it, we would still call the infection a disease. These
points, plus Goosens' example of hideous death in the service of
reproduction or population control, show that dysfunction is not
necessary for disease. Dysfunction is not sufficient because a
viral infection that interfered with the self-destruct aging function would not count for us as disease.

My answer to Goosens and Reznek is as follows. Precisely
because of the point Goosens stresses—the ubiquity of sacrifice

by parents in reproduction—I would expect the disease concept to prefer reproduction over survival even in these hypothetical cases. Normal pregnancy and birth already put considerable stress on women's bodies, not to mention the high morbidity and mortality from both without medical treatment. If aging, or even horrible death, had a reproductive function, then I do not see how it could properly be called a disease and, contrary to these writers, what interfered with it could indeed be called a disease.[72] Surely Goosens gives the game away with his slug example. It is wholly impossible for a normal stage of development to be a disease or pathological. Of course turning into a slug to reproduce is not a disease, if that is the way your species reproduces. To call normal reproduction pathological is to explode the disease concept entirely. And, notably, Goosens shifts from "disease" to talking of "medically significant condition," a category into which he then classifies ordinary pregnancy. Reznek sticks to disease and pathological condition, but I think he is mistaken.[73]

Obviously, these are cases where we might passionately want to be abnormal, and might invoke medical treatment to acquire or protect our abnormality. But Goosens and Reznek confuse desirability or treatability with normality. The fact that reproduction bears a horrible cost does not mean that its typical processes are pathological. Rather, what one should conclude is that one prefers not to reproduce, and may enlist medicine's help in avoiding doing so. But that is precisely the present situation. That is, many doctors prescribe contraceptives, tie off Fallopian tubes, and so forth, precisely because many women find it important to avoid present or future reproduction. Goosens' hypothetical cases are more dramatic, but I do not see that they change the basic point he began with: normal reproduction often carries a very high cost to the individual. But that cannot make normal reproduction pathological, or a disease.[74] And, of course, I hope that doctors will guard women's orgasms against infection even if they have no evolutionary function.[75]

Clinical Implications

Wulff et al. (1986), like some other writers, see the BST as implying an inadequate and insensitive approach to clinical medicine. Concluding their dialogue over BST-like theories, they write:

> [W]e reject biological reductionism in medicine. ... We accept that the mechanical model is an indispensable *part* of the disease concept and we only object to the contention that it offers a *complete* description of disease.... [D]iseases are not only biological entities. It is not biological organisms, but human beings, who are ill, and even diseases, like a duodenal ulcer or a cancer, which clearly involve a biological defect, have causes, manifestations and effects which reach far beyond the limits of biology. Therefore, clinical medicine is more than applied biology. Clinicians must also take into account their patients' experience of pain, suffering, self-respect, aim in life, etc., and they must learn to deal with such non-biological phenomena in a rational way. ... At best, the reduction of non-biological phenomena to biology is futile, and, at worst, it leads to a distorted and unacceptable view of man. (Wulff, 1986, pp. 58,59)

As a "wider frame" for a "more adequate concept of man," the authors propose a "hermeneutic" one.

Previously, to illustrate the points in the passage just quoted, Wulff et al. had given four clinical examples. Three were duodenal-ulcer cases in which the pathology and drug treatment were identical, but "the meaning which these patients assigned to their illness was vastly different" (p. 56). One man had had ulcers before, hence was undisturbed, confident the drug would remove his symptoms. A second had never been ill and was terrified his ulcer was as lethal as his dead brother's cancer. The third, who attributed his ulcer to stress, was embarrassed and fearful of the diagnosis' effects on his job prospects. The authors also cite Cassell's case history of a 35-year-old sculptress with metastasizing breast cancer. Although this woman's symptoms and the side-effects of her treatment were unpleasant enough, the cancer's real significance for her was its threat to her lifespan and to her identity as artist, friend, and independent woman. Thus, to understand these four patients' diseases, one must "take into account all the aspects of personhood," not just their "mechanical fault" or "biological dysfunction" (pp. 56,57).

This criticism is easy to answer. Almost nothing in Wulff et al.'s views seems to me to contradict the BST. To begin with, some of the appearance of conflict comes from the authors' too-limited view of biology. I cannot agree that "it is not biological organisms, but human beings, who are ill," since human beings are biological organisms, i.e., they are nothing but one species of fauna evolved on Earth. But for that very reason, biology must include psychology, which is why the BST applies to mental health. So the "causes, manifestations, and effects" of disease that the authors say "reach far beyond the limits of biology" do not necessarily do so on a proper view of biology. The authors' debate over whether disease is more than biological dysfunction may be partly a terminological one. Conceptually, I prefer to separate a disease (ulcer or cancer) from the patient's thoughts and fears about it (loss of job or identity as independent woman). But surely Wulff et al. and I can make all the same points in different language, and take the same view of good clinical medicine. Where they say the sculptress' disease includes her loss of self-esteem, I say it causes her loss of self-esteem (unless that is itself psychopathology), But we all agree that a good doctor must be sensitive to "the meaning she assigns to her illness." And this is so not only because doctors should treat their patients humanely as suffering people—already an important point—but also because psychology often causally affects disease. Contrary to the impression left by some critics, nothing in the BST precludes psychogenic causation. For all these reasons, I wholly agree that in Wulff et al.'s limited sense of biology as physiology or pathology, "clinical medicine is more than applied biology."

Similarly, I was pained to hear of a college course presenting my views as entailing a callous approach to the following case (Kleinman, 1988, p. 131 ff). A middle-aged black woman, deserted by two husbands, living in a slum in poverty with four troublesome children and a disabled mother, develops hypertension. What she believes herself to have is a folk disease that her community calls "high blood," wherein blood, by accumulating too high up in her body, causes various ill effects requiring certain folk remedies, including a salt-rich diet. What does the BST imply about this case? First, it does not imply that the woman's hypertension is not caused, wholly or partly, by stress or other psychological factors. That is a straightforward scientific issue, though not, I think, a hermeneutic one. So the BST in no way implies that the most effective way to cure this woman's or other people's diseases might not be to improve their neighborhood, income, or marital status. Of course, physicians often lack any special wisdom or power to change these socioeconomic factors, except by identifying them in counseling the patient.

On the other hand, the BST does imply that there is no such disease as "high blood"; this folk theory describes imaginary phenomena. This woman is like a 19th-century masturbation theorist in having false beliefs about her disease process and its proper treatment. However, that need not keep a physician from treating her sympathetically and respectfully, beliefs and all. Even among wealthy and well-educated Americans, false beliefs about health, disease, and therapy are rampant. Many upscale patients have a quasi-religious faith in vitamins, minerals, amino acids, hormones, and other scientific objects that resembles a biomedical cargo cult. How to deal with patient ignorance or irrationality is a clinical problem on which the BST has nothing to say, except to point out the obvious, that patients' theories of disease and therapy are not always true.[76]

In short, the BST is compatible with various views about clinical medicine, including most of what authors like Wulff et al., vdS/T, and Fulford have to say. Their clinical versions of the

BST are, I think, straw men based on confusing theoretical and practical health. Thus, the BST can grant that the personal significance of the same disease differs greatly among patients. It can grant that clinical medicine is a species of human relations, and therefore art as well as science. It can recognize multifactorial disease causation, including psychological factors that may themselves depend on a social, economic, or cultural environment. (Of course, Wulff et al. and vdS/T might not agree with me on the proper form of psychological science.) It leaves room for almost every imaginable controversy in medical ethics. The BST does insist that all genuine disease or illness must involve biological dysfunction, on the broad view of biology as including psychology. But that merely implies that not every problem with which a patient comes to a doctor's office, or leaves it, is a problem of health. And one of Wulff et al.'s own examples makes a related point in any case. They suggest that veterinary euthanasia shows that "there is more to life, even animal life, than survival and reproduction" (p. 56). What it really shows is simply that doctors can be justified in doing things besides promoting health. (Perhaps some normativists would say that euthanasia improves a patient's health, but that is absurd; one can be better off dead, but not healthier.)

If and when it is right for a doctor humanely to kill a terminally ill cat or human being, that shows what is true in any case: that doctors are sometimes justified in serving other values than health. We must simply bear in mind that their nontherapeutic efforts are usually more controversial than their therapeutic ones. The practical importance of a clear definition of disease, to me, is that it marks out a domain in which there is a very strong presumption of disvalue. The disvalue of disease and the benefit of treating it tend to be barely controversial for at least three reasons:

1. The general excellence of biological designs in serving survival, a goal of almost everyone;
2. The fact that biological functions usually harmonize with one another, so that one dysfunction leads to another; and

3. The neutrality of biological functions to most choices of activity or lifestyle.

Thus, the practical importance of calling something a disease is that such status can almost always be presumed to answer the question of its disvalue and suitability for therapy. But it is only a presumption: I have emphasized that there can be diseases that are neither disvaluable nor worthy of therapy. Conversely, physicians can be justified in nontherapeutic activities. So the concepts of health and disease are far from settling all clinical or social questions. The main thing is to avoid false presumptions caused by calling something a disease (e.g., masturbation) which lacks the biological dysfunction on which alone such presumptions depend. But that is just what has often occurred in psychiatry and even somatic medicine.

Conclusion

With apologies for its length, I hope that my discussion supports the following conclusions. First, the BST easily handles most of the objections and counterexamples posed by critics. A few more challenging cases seem, at worst, to need further analysis. A full defense of the BST requires that one canvass competing accounts of health to show that they either do no better with these cases, or do better at too high a cost. While this would require a separate paper, we saw several examples of the high price some writers pay for their normativism about health. Some call pregnancy, wanted or unwanted, a disease. Some are normativists about science in general, or biology, or biological function. Some are normativists about specific diagnoses (tuberculosis) or humdrum physiological categories (kidney). And no normativist theory is really univocal in defining health for human beings, animals, and plants. Some readers will find these problems grave enough to prefer to tolerate the BST's small annoyances, such as vagueness in environmental disease or a novel view of lice.

Second, some BST critics are attacking the concept of disease itself, not my analysis of it. That is most clearly true of Engelhardt, but arguably also of others who offer adaptationist theories (Bechtel, Reznek) or positive theories (Ladd) of health. Whether health that is more than the absence of disease can really be health, or must instead be something else entirely, is an important issue dividing me from many critics. I do see room for fruitful debate over individual, nontypological, environmentally relative, or positive ideas of health.

Third, a sign of progress is that one of the best recent papers (Wakefield, 1992a), by analyzing disease as harmful dysfunction, covers almost the whole gap separating the BST from major rivals. As soon as biological dysfunction is required for disorder, virtually all the BST's benefits accrue in clarifying professional and social controversies and preventing political abuse of medical vocabulary. The BST's difference from Wakefield is theoretically important but at present of limited practical importance to psychiatry.

Finally, I hope my discussion shows how the BST offers the best of both analytic worlds. It provides a theoretical, value-free concept of disease or pathological condition. But on this foundation one can build value-laden disease concepts, by adding evaluative criteria, to taste. Starting from the basic disease concept, one can define clinically evident disease, or harmful disease, or serious disease, or treatable disease, or disabling disease, or disease that should be covered by insurance, or disease that should remove civil or criminal responsibility, and so on. Best of all, one can use different "disease-plus" concepts for different purposes. Yet the value-free scientific disease concept remains as a bedrock requirement to block the subversion of medicine by political rhetoric or normative eccentricity. Fulford concludes in his Chapter 3 debate that

> so far as medicine is concerned, ... non-descriptivism provides a far richer theoretical resource than descriptivism

[e.g., to Fulford, the BST] for dealing with the ethical problems of clinical practice. (Fulford, 1989, p. 54).

But that could not possibly be true, since the BST can simply add
on any of Fulford's normative "resources" to resolve clinical or
social issues. That is the beauty of a multilevel approach, which
separates the scientific foundation from a variety of normative
superstructures that can be erected upon it. When this virtue of
the BST is appreciated—its practical flexibility and conceptual
richness—perhaps it will find more converts in the next two
decades than it has in the two preceding.

Acknowledgments

I thank Roy Sorensen, Mary Williams, and Karen Rosenberg for ideas.

Notes

1. The original group consisted of one independent essay
   on functions (WF) and three on health (DI, MH, HTC)
   that presupposed it (for abbreviations, *see* references).
   These three came from one long manuscript (cited by
   Margolis, 1976) that I divided for ease of publication.
   Unfortunately, this division was inconvenient for
   critics, some of whom never found all four papers. My
   chapter CH summarizes the BST and also modifies and
   extends it a little. I make some application of the BST
   to a concrete controversy in PMS (a paper written in
   law-review format but published in another, thus annoying to read).
   Pawelzik (1990, p. 18) raises premenstrual syndrome
   as an objection to the BST. He approves DSM-III-R's
   new category of "premenstrual dysphoric disorder," but
   suggests that my analysis mishandles PMS. In fact,

with this diagnostic category the APA took exactly my approach, to separate normal PMS from a pathologically extreme version (*see* DSM-IV, p. 715).

2. A word on terminology: My position has been variously named in the literature. Although I christened the general thesis that health is value-laden "normativism," I did not dub its opposite. Since I used "descriptive" as the antonym of "normative," the natural noun would have been "descriptivism." But later writers instead called my general thesis "naturalism," "neutralism," or "objectivism." Both "naturalism" and "descriptivism" risk confusion because they are standard terms for a doctrine about evaluative concepts, a class to which I deny health belongs; *see* discussion of Fulford below. My specific naturalist theory has been called "functionalism" and several other names, but I will adopt Nordenfelt's concise coinage.

3. However, my literature search was based only on the three standard citation indexes. I apologize to any critics I have overlooked or slighted, including the authors of a handful of citations to which I did not gain access before publication. And I regret having translated Pawelzik's essay (1990) too late to give it the coverage it deserves.

4. This definition from HTC (pp. 562, 567) is more exact than the more often quoted summaries of it in DI (p. 59) and MH (pp. 62, 63).

5. Reznek (1987), p. 124. He also cites the steatopygy of pygmy women and the unusual stomachs of certain aborigines as race-specific adaptations for food and water storage, respectively. He gives an example of a tribe-specific adaptation (urine concentration by the kidney), but refuses to relativize normal functioning further to a tribe. His only explanation is that tribes interbreed—but since races do too, I do not quite grasp his position.

6. Another relevant point about functions is that Bigelow and Pargetter's (1987) "propensity" interpretation of fit-

ness, though often seen as a way to enhance a Wrightian etiological analysis, was originally joined to a goal-analysis like mine. Thus, one can have the BST with etiological functions, propensity-based or not, or with goal-functions, propensity-based or not. But I take no responsibility for counterexamples to the BST that presuppose an etiological function theory.

7. *See*, for example, Symons (1979), Buss (1994), and the references in Wakefield (1992a). A related new area of interest is evolutionary medicine (Nesse and Williams, 1995).

8. *See* CH 379–382 and references therein; Wakefield (1992a,b), (1993).

9. Bechtel, who only cites DI, does not seem to have looked at HTC or WF, which precisely try to meet what he calls the "challenge in the case of nonartifacts ... to ascertain what the design is" (p. 144). His own discussion of teleology first considers Wright's theory of function, then uses a point by Margolis to introduce the propensity interpretation of fitness. That point allows us to "reverse [Wright's] connection between selection and function" (p. 151), i.e., to "claim that those things that are functional will evolve, rather than to claim that those things that evolved are functional" (p. 150). Since this is just my own criticism of Wright, as far as I can tell Bechtel's final view of functions is consistent with a propensity version of my own goal-analysis in WF. What I cannot accept is Bechtel's subsequent adaptationist view of health. I disagree already with his claims that physiology is about homeostasis (pp. 148,149), and that to improve the "physiological constitution ... that past evolution has provided" in a way that raises fitness is to improve health (p. 151).

10. Ruse (1981) seems to assume that the medical concepts of health and disease have perfect precision. He concludes (p. 719) that the BST "needs revision" to "remain plau-

sible," because on several theories he discusses, the BST's implications for homosexuality seem unclear. But difficulty in classifying a condition is an objection to the BST only if medicine's own view of that condition is clear. Ruse offers no evidence that medicine could clearly classify homosexuality on these theories either.

11. His reason is that the definition of "kidney" is "organ with the function of excreting waste products," and the term "waste" is evaluative (pp. 51, 52). This is a good example of a common phenomenon: normativists about health being implausibly normativist about other things too. Many people hate rats, and so presumably wish that all rats would go into renal failure, retain everything, and die. On Fulford's view, for such people there is no such thing as waste in a rat—it's all excellent stuff—so they must believe that rats have no kidneys, which is absurd. (A more modest view would be that in rats, "renal failure" is actually renal success.) Up to a point, one can avoid this objection by distinguishing what is good for rats from what is good *tout court*. Even rat-haters may concede that it is in rats' interest, though not in ours, for them to stay alive by excreting. But what if we instead hate a nonsentient animal or plant—wasps, fleas, or kudzu? Many philosophers maintain that nonsentient beings can have no interests. Then must we say that these organisms, unlike sentient ones and nonsentient ones that we happen to like, have no mechanism for excreting waste? In reality, "waste" is a perfect example of the kind of term that physiologists consider value-free, and rightly so in a Sommerhoff-style framework. So the example is relevant to the points in the text.

Incidentally, my own moral view on the interests or welfare of plants and lower animals is that nonsentient beings have neither. What they have is goals. But promoting their goals is not in their interests, since they have none. One can only speak of what is good for a plant in the biological sense of what makes it flourish, i.e., vigorously pursue its goals. But this use of "good for" is close to metaphoric; it is no part of a moral theory of value. Here Sommerhoff's concept of goal-directedness helps clarify an important issue in ethics, the moral standing of nonhuman organisms.

12. Cf. HTC, p. 559; CH, p. 371. The point that "function" places its own sign on statistics also answers Engelhardt's charge (1976b, p. 265) that we need value judgments to exclude statistically supernormal abilities like unusual athletic talent from the realm of disease. On my analysis of function, "there is no such thing as excessive function" (HTC, p. 559).

13. But this is no objection for Fulford, who with many other normativists is content for biology to be value-laden (*see* his Chapter 6).

14. In short, I see what Fulford quotes as harmless pieces of rhetoric no more confusing, say, than financial newsreaders reporting that IBM today is "better by 5/8." Although "better" is a quintessential evaluative term, here it is just a colorful variant of "higher." No one takes it as a sincere condemnation of short-sellers or future buyers.

15. For example, this point is missed in a review of Fulford's book. Pickering says that the Chapter-3 debate features a "half hearted" descriptivist (1991, p. 180). I thank him for noting that my views are largely absent from the debate, and also that HTC answers the kidney point (p. 180). But it could not possibly be a "descriptivist" about values who represented me in the first place.

16. Cf. Goosens' criticism (1980, p. 101) of Whitbeck (1978) for claiming to be a normativist, yet offering definitions of health and disease that mention values rather than using them.

17. Fulford (1989) p. 55. On p. 53, he concedes that the tendency of "descriptivists" to ignore "non-empirical" prob-

lems in medicine is only a tendency. However, he thinks
I "illustrate" it "rather well," since I want to "disentangl[e]
medicine *from,* rather than to resolv[e], ... social and ethical issues. ... " I do not see how that judgment can be
passed on DI, MH, or CH. At any rate, to me, separating
theoretical issues about pathology from evaluative issues
greatly sharpens one's evaluative focus.

18. *See also* Agich (1983, p. 31), who repeats Engelhardt's
criticism, also using the sickle case. Noting that species
often go extinct, Agich adds that "Boorse's preference for
survival amounts to no more than an unjustified value
stipulation in terms of which 'disease' and theoretical
health are defined."

19. CH, p. 372. To clear up an inconsistency, my first point
here concerned Engelhardt's assumption that biological
functions must aim at species rather than individual survival (since he attacks the BST for preferring the former
to the latter). This premise could only come from a popular misconception of evolution as organized around species survival. Instead, if George Williams (1966) was right
about group selection, then no traits are selected specifically for their effect on the species, rather than on the
individual and his kin. In that case, no physiological
functions will aim at species survival directly. But, as
HTC noted (p. 556), on Sommerhoff's analysis physiological functions are also directed at species survival
whenever they are directed at individual survival. Whatever causes individual survival helps cause species survival *a fortiori,* since a species survives as long as at least
one member does.

20. W. Miller Brown (1985), pp. 315, 316. *See also*
Engelhardt's (1986), pp. 167–171, which makes similar
points without the misattribution on species survival.

21. Besides its evaluative features, Engelhardt also believes
the disease concept has descriptive, causal-explanatory,

and social functions (1975, 1976a,b, 1986). However, I do not think these other features affect my point in the text.

22. Admittedly, in the previous two pages Engelhardt suggested that only what a whole society "accepts" as a clinical problem can be called a disease. He said that "the treatment of diseases is a societal undertaking" and that "the term *disease* [has] connotations of an objective reality or a general intersubjective agreement." Thus, the term might be "inappropriate" for homosexuality or pregnancy, which "only certain persons, in certain circumstances, may see as clinical problems to be treated" (pp. 172,173). In other words, he hints at an analysis of disease as "consensus clinical problem." Since, as shown earlier, he himself employs an environment-relative concept of disease (the black in Trondheim), it is hard to see why he here resists a person-relative one ("disease for X"). In any case, a consensus-clinical-problem account of disease still implies that fertility or pregnancy would be diseases if all the women in a society (e.g., China) disvalued them. But that is absurd. Already we have social consensus on the undesirability of unwanted pregnancy in many situations (environments?)—extreme youth, poverty, rape, and so on. Yet such unwanted pregnancies are not called diseases by any medical source I know. Another counterexample to Engelhardt's version of the medical disease concept is menstruation, which virtually all women do disvalue (and which is medically suppressible by progesterone), but which is normal in the eyes of medicine. The best counterexample of all is male circumcision.

23. To forestall a confusion, HTC assumed—e.g., in referring to the right amount of thyroxine for "current metabolic needs" (p. 559), or "typical occasions" (p. 562)—that normal physiology is a "dynamic equilibrium." *See* Dynamic Equilibrium and Environmental Disease for discussion.

24. Note that absent the interbreeding criterion of Mayr's "biological species concept"—for example, in paleontology—one effect of the antitypological critique is to diminish the objectivity of species boundaries. This suggests a quick argument that vdS/T's antitypological points do not threaten the medical usefulness of a concept of species design. Never, as far as I know, is medicine in any doubt whether an organism belongs to our species, even without an interbreeding test.

25. Sober (1984), p. 88 ff ("resistance to disease" and "epidemic"); Williams (1992), index and p. 27 ("disease resistance"); Hull (1974), p. 66 ("disease"); Mayr (1988), p. 106 ("pathogen"); Lewontin (1970), p. 15 ("pathogen"); and so forth.

26. Note that both Engelhardt's and Margolis's views on disease make it difficult or impossible for animals and plants literally to be diseased. Margolis (1976) ultimately grounds disease status on a condition's relation to "the prudential interests of the race" (p. 253). Such a theory must be "adjusted" for lower animals "by extension" and for plants "by analogy" (p. 252)—a distant analogy indeed if plants have no prudential interests.

   Engelhardt is still worse off regarding nonhuman disease, since he wants ultimately to view human disease as interfering with rational free agency (1976b, p. 266), a Kantian concept specifically designed not to apply to other animals, let alone plants. So he must emphatically adopt Sedgwick's view (1973, p. 31) that

   the paradigm examples of diseases in animals and plants are those that afflict pets, or animals and plants grown for food, and so on. It is difficult to speak of feral animals and plants as being diseased (or at least diseased in a clinical sense) except by analogy. (Engelhardt, 1986, p. 198)

   Conceding that the BST, by contrast, applies routinely to wild animals and plants, he complains that such a concept fails to focus on the sufferings of animals. The concept of disease held by veterinarians or agricultural scientists is, in contrast, dependent on what humans hold to be the proper functions, levels of pain, and characters of grace and form for particular groups of living entities. (Engelhardt, 1986, p. 198)

This claim about the usage of veterinarians and plant pathologists is, I believe, almost as clearly false as his claims about medical usage. Veterinarians do not, for example, consider pathological the failure of a dog to match breeders' criteria of "grace and form" for its breed. A really awful Lhasa Apso, a dog miles away from the Lhasa Apso ideal, is not thereby diseased; otherwise mongrels would be masses of pathology. In reality, mongrels are healthier than pure-breds, so here exquisitely detailed human standards of "grace and form" collide with medical standards of health. Similarly, if what a farmer intended as a lard pig turns out more like a pork pig or bacon pig, veterinarians do not consider this commercial mismatch pathological, however upset the farmer may be. Conversely, biologists speak of diseases and pathogens in "feral animals and plants" without any sense of oddity or of commitment to relief of plant "suffering," whatever that might be. On Engelhardt's view, the book *Diseases in Free-Living Wild Animals* (McDiarmid, 1969) should not exist, nor the *Journal of Wildlife Diseases*. It is, of course, trivially true that it would be odd to speak of species that never end up in a clinic as "clinically diseased." But those that do are, like humans, clinically diseased if they are diseased in a way that can be diagnosed and/or should be treated, i.e., have a diagnostic or therapeutic abnormality. So it seems to be the BST that handles all our vocabulary of nonhuman disease both pathological and clinical.

Similar problems affect Nordenfelt's and Reznek's accounts. Quite plausibly, Nordenfelt (1987, p. 5) makes it an adequacy condition on a health theory that it explain plant and animal health as "*not* radically different" from human health. Yet it is hard to see how his own theory meets that test. Reznek (1987, pp. 135,165) cites von Wright's idea that all living beings "have a good or well-being which can either be promoted or impaired." But I fail to see how any vegetable good is remotely analogous to Reznek's extremely complex analysis of human good (pp. 140–152) in terms of "worthwhile desires and pleasures." Among theories in the field, only naturalist ones offer a univocal concept of disease for humans, animals, and plants.

27. Regarding individual development, however, Sober does virtually hold the position that "natural" is an unbiological term. Here he is basing his discussion on Lewontin's analysis of the "norm of reaction," the function that shows how a given genotype produces different phenotypes in different environments. An example is "the height that a corn plant with a particular genotype will attain as a function of how much water or nutrition or sunlight it receives" (Sober, 1984, p. 160). In drawing the moral of modern biology, Sober then makes two claims, one modest and one sweeping: that there is no one uniquely natural phenotype; and that all possible phenotypes are [equally] natural results of development. He says of the corn plant not only that "[t]here is no such thing as its 'natural' height," but also that "[a]ll possible phenotypes of a genotype are 'natural,' since all are possible" (p. 160; cf. 1994, pp. 223, 234). (Thus, though Sober does not say "equally," it is implied and his discussion makes little sense without it.) Following Lewontin, Sober then applies this point to children's IQ.

Both medicine and the BST can accept the first claim. But neither can accept the second, which instead marks a sharp contrast between medical thought and Sober's version of contemporary biology. In my opinion, that is because medicine is right and Sober is wrong. It is obviously false that there is no interesting or fundamental biological distinction among outcomes of development. As Sober notes (1992, p. 223), in some environments the corn plant or child dies. No distinction is more central to biology than life vs death. Even construing naturalness causally, as Sober does, it is at best misleading to say that every outcome is equally caused by organism and environment. If one pours gasoline on a corn plant or child and ignites it, the organism dies, but its only contribution to that outcome is to be vulnerable to 1000°C temperatures. In burning, it is expressing only that tiny part of its nature which it shares with all organisms, indeed most substances, on earth. After incineration, as with any other dead animal, none of its genetic information, Sommerhoff goal-directedness, or species or individual characteristics are expressed at all. Since I see no biological sense in which a dead species member could be as normal as a live one, I think much of what Sober says about development is either trivial or false, and where false, has appalling implications for children. But here I will simply admit that insofar as it is typological or essentialist to deny that live and dead organisms are equally normal, both medicine and the BST are incontestably so.

28. What he actually says is that being the wrong race could be a "medical problem" (Engelhardt, 1986, p. 168) or "problem for medicine" (p. 169). But as we have seen, he does not separate consensus medical problems from diseases. Moreover, the race examples appear in the midst of his critique of the BST, to which they are relevant only if Engelhardt means to call them diseases.

29. Admittedly, some authors narrow their health concept to fit a narrow concept of disease, or even illness. For example, Ladd (1982, p. 29) and Downie and Telfer (1980, p. 20) agree that a blind, deaf person with no arms or legs

can be in perfect health. But to narrow health in this way is essentially to change the subject from the health concept I am analyzing. On the narrow usage, one can no longer describe medicine and allied fields as "health sciences," normal medical coverage as "health insurance," and so on. An employee would be astonished to discover that his "health insurance" excluded injuries, poisonings, and so on.

30. Similarly, Goosens (1980, p. 109), after shifting from "disease" to "medically relevant condition," includes pregnancy in the latter category. But by this maneuver he changes the subject.

31. Of course, the same reading may be diagnostically significant only for one, as revealing some other underlying disease. An athlete who suddenly acquires high-normal blood pressure might have cause for grave concern. That is Aubrey Lewis's point in the passage Brown quotes: "in clinical practice the physician must take the patient pretty much as supplying his own norm of total performance or behaviour, and proceed by rough and ready appraisal of whether there has been any departure from this" (1955, p. 113). But that is a diagnostic point. Diagnostic and therapeutic points in no way show that a given disease can be only pathological for athletes. No one supplies his own norm of theoretical health.

32. Margolis (1976), pp. 245, 243. Margolis describes his analysis as only "a first approximation to the theory of medical norms," but in his paper does not appear to alter the views I quote.

33. Agich distinguishes between "ill" and "sick." To be ill is simply to "feel bad" or have "discomfort," whereas to be sick is to be ill and socially assigned to the Parsonian "sick role" (Agich, 1983, p. 35). The former thesis seems to me remarkably overbroad, making illness out of all life's disappointments. For example, it implies that Agich's and

every other critic's paper made me ill by rejecting the BST. I do not see any warrant in ordinary language for distinguishing "sick" and "ill," so whatever value there is in Parsons' concept, it should apply equally to either.

34. One of countless age-related examples of accumulating pathology is the Achilles tendon. Although extremely strong, it is poorly vascularized, especially in the thinnest ("watershed") area. As its blood supply progressively diminishes in middle age, stress-caused microtears heal with scar tissue, a clinically silent process. In time, the result is a much-scarred tendon liable to rupture under stress without warning.

35. HTC 561, a page that contains a full answer to the problems discussed in the text. Wakefield (1992a), focusing on reproduction, says that:

A condition can reduce fertility without causing real harm; marginally lowered fertility is serious over the evolutionary time scale, but it may not affect an individual's well-being if the capacity for bearing some children remains intact. And some serious harms, such as chronic pain or loss of pleasure, might not reduce fertility or longevity at all; Kendell (1975) admitted there are many harmful physical conditions, such as postherpetic neuralgia and psoriasis, that are clear cases of disorder but have no effect on mortality or fertility. (p. 378)

However, Wakefield uses this point to show that "the harm requirement must be added to, rather than derived from, the evolutionary requirement," and that is his reason for rejecting the "biological-disadvantage" approach supposedly including Scadding, Kendell, and me. Wakefield's logic here is faulty or his criticism disingenuous. Since he himself later requires dysfunction for disorder and adopts Wright's theory of function, if he accepts psoriasis as a clear case of disorder, he accepts that it involves dysfunction, hence statistical effects on longevity or fertility. Since the BST, like Wakefield but unlike Scadding and Kendell, is based on dysfunction, Wakefield has here posed no problem for me. It is misleading to group me with Kendell and Scadding, saying, for example, that I claimed that "a disorder is a condition that reduces longevity and fertility." That overlooks the difference noted by Scadding himself: my dysfunction requirement.

Wakefield's fine papers incorporate much of the philosophical progress made on these topics and apply it to contemporary psychiatric issues. But he sometimes leaves unclear the relation of his views to others', especially Wright's and mine. Wakefield's final analysis of disorder is essentially the BST with Wright's view of functions, plus what I view as an unnecessary added harm requirement (as in Reznek).

36. Note, however, that clinicians often seem to use "disease" to mean "clinical disease." Thus, they may declare a patient to have "no disease" when what they mean is "no clinical disease." Here and elsewhere in this section, the difference between pathologists' and clinicians' use of the term "disease" recalls the pattern of shifting precision levels in the language-game discussed by David Lewis (1979). My claim is that the pathological level of analysis is the conceptually fundamental one in scientific medicine: "clinical disease" is definable via "pathological," but not conversely.

37. Miller Brown 1985, p. 326. For further debate over medicine as science, *see* Munson (1981), a paper with almost all of which I agree. Munson avows his support of a naturalism based on biological function (p. 201) but rejects my account of function for unspecified reasons (p. 207n).

38. Pathology's organization around the distinctive concept of disease is one disanalogy between it and electronics,

which involves no comparable new scientific category that I know of.

39. *A fortiori,* I feel entitled to reject individual medical sources' disease judgments if there is reason to doubt they report a consensus. Thus, Pawelzik (1990, p. 31n) finds the *Cecil Textbook of Medicine* saying: "Scurvy is an inborn error of metabolism ... kept in remission by vitamin C" (18th ed., p. 146). This remark, though charming, hardly fits the view of scurvy in medical dictionaries or general usage. For another example, *see* Note 57.

40. Engelhardt's text here raises the question whether he sees any distinction between a negatively evaluative term ("trash," "weed") and a value-neutral term for a disvalued object ("feces," "vomit," "death"). He writes:

> Often the values lie hidden within the disease term itself, as occurs in "George is a schizophrenic." Schizophrenia is not valued; to be judged schizophrenic is to bear a disvalue. Whether it is athlete's foot, tuberculosis, a deformed nose, or unwanted pregnancy, diseases or clinical problems are not good things to have. They are things that are good to prevent, treat, or cure. (Engelhardt, 1986, pp. 174, 175)

This passage seems to be meant as argument. Yet his latter statements ("Diseases are not good to have," and so on) everyone concedes to be usually true. The issue is whether they are always true, i.e., whether diseases are always bad to have, and whether that evaluation is part of ("hidden within") the meaning of "disease" itself. As vdS/T put it: Does the "is" in statements like "Health is desirable" state an empirical or conceptual connection (1988, p. 80)?

Engelhardt's normativism about health would be more convincing if it were clearer that he grasps these distinctions. Here is an earlier passage raising the same doubts:

One should note the therapeutic injunction implicit as the bias of all illness and disease language. That is, to recognize that one is ill or diseased implies that one ought, *ceteris paribus,* do something about it. Admitting to being ill or diseased usually invites one to explain if one does not seek treatment. Compare, for example, the following sentences:

> I feel ill but don't care.
> I'm glad I feel ill.
> I feel ill but do not wish to seek treatment.
> I wish I felt ill.
> I have a disease but don't care.
> I'm glad I have a disease.
> I have a serious curable disease, but do not wish to seek treatment.
> I wish I had a serious disease.
> I wish I had a disease so that I could avoid the draft.
> I wish I were disabled and could draw the full disability payment stipulated in the company's insurance policy.

Only the last two sentences do not invite further explanations in order not to seem senseless. (1980, pp. 248, 249)

As an argument for normativism about disease or illness, this passage has no force, since it is easy to construct a parallel series of equally puzzling remarks using clearly descriptive vocabulary:

> My fingers are falling off, but I don't care.
> I'm glad this ship is sinking.
> I feel pain in each of my 206 bones but do not wish to seek treatment.
> I wish I were being torn apart by starving junkyard dogs.
> I'm hanging by my testicles, but I don't care.
> I'm glad I eat only cat vomit.
> All my muscles will soon be paralyzed, but I do not wish to seek treatment.
> I wish everyone I love would be sliced into pieces by Iraqis.

All these statements, to say the least, "invite further explanations." Thus, Engelhardt's own examples allow the possibility that "disease" and "illness" are descriptive terms for things that are (like pain, death, and cat vomit) generally disvalued.

41. Wulff et al. (1986, pp. 52, 53) also offer an etymological argument against the BST.

42. Of course, this analogy is misleading if heat and cold are natural kinds, but health and disease are not. That cold is only the absence of heat is a discovery. I suspect, however, that equally, medical science discovered that the real essence of experienced illness is biological dysfunction (disease). In any case, non-natural-kind examples could easily be given.

43. An object can be heavy or light in weight, large or small in size, young or old in age. But "weighty," "sizeable," and "aged" refer only to one end of the spectrum.

44. Fulford (1989), pp. 44, 45. Contrary to Fulford's remarks on "functioning as a race-car driver" (p. 45), I stressed in DI (p. 58) that in health contexts, "function" is to be predicated of parts or processes in organisms, not of the whole organism as in Aristotle.

45. Fulford (1989), Chapter 2, refers extensively to "ordinary usage" not only of "illness," but also of "disease" and "dysfunction." But in Chapter 3, he concentrates more on "technical medical usage" (p. 35 ff). *See also* Reznek (1987), p. 67.

46. Once I had the experience of debating the BST with a friend, a very intelligent philosopher, who immediately abandoned his line of criticism upon realizing that he was rejecting diabetes mellitus, cancer, atherosclerosis, and other leading causes of death as diseases because his intuitions tacitly assumed that all diseases are infectious.

47. Among hypotheticals, one should also place more trust in those more conservative of our stock of beliefs. The more

extreme a case-world's divergence from our world, the less we can trust our intuitions about what we would say. Cf. Fodor (1964).

48. On disease classification *see* Reznek (1987), chapters 10 and 11.

49. I would take a similar line on Margolis' point that the "onset" of a disease can precede dysfunction (Margolis, 1976, p. 243). Such a condition will be disease, but isn't yet. I do not think his example, cancer, works since cancer cells not only have internal dysfunction, but, being malignant, also invasively kill neighboring cells. (But he may be thinking of precancerous conditions like cervical dysplasia or certain moles.) *See also* Pawelzik (1990), p. 16. Reznek cites a benign tumor without dysfunction. The view I have taken (HTC, p. 565) is that this is a purely structural condition which medical books call pathological, but should not.

50. Engelhardt (1986), p. 169. Goosens (1980, p. 113) likewise argues that basing health on functions makes the past improperly relevant to what is a disease.

51. To prevent confusion: My analysis of function, unlike Wright's, requires that a function actually be performed at the time to which the statement refers. But in calling a function species-typical, one generalizes over a reference class of species members, including past ones. That is why a species-typical function of a body part may not actually be performed in an individual. All functions are actual effects, but some species-typical effects may occur only in a species subset to which the individual does not belong, e.g., a past subset.

52. By no means, however, do I share the fashionable view that aggression should or can be eliminated from human psychology. Aggression is an essential force without which human beings would achieve very little and resemble us so little as to be barely human. In an era of genetic tech-

nology, there is scarcely a more dangerous delusion than some writers' dream of making simple, revolutionary improvements in the human psychological design. In my opinion, we are about as likely to be able to do this as a chimpanzee is to design Intel's next microchip.

53. Interestingly, medicine does not always follow the physiologist's clear concept of penetration: crossing a living membrane. To the physiologist food, or penises, do not actually penetrate the body. Your dinner is not inside you until it is digested and absorbed across the intestinal mucosa. Like the air in your lungs, it is merely against your interior surface. Yet medical sources list foreign objects in the stomach as pathological. Among worms, roundworms (as in trichinellosis) and some tapeworm larvae are inside tissues like muscle. But adult *(Taenia solium)* tapeworms simply attach themselves to the intestinal wall, though the attachment does some slight mucosal damage and can provoke irritation and eosinophilia. Similarly, cholera bacteria are on, not in, the mucosa. By contrast with most infectious diseases, they do their damage by producing enterotoxins that cause the small bowel mucosa to pour out its electrolytes, so that the patient dies from massive fluid and electrolyte loss. But in view of my next remarks on hanging, this point hardly threatens the medical classification of cholera, tapeworm, or even foreign bodies. The internal effects of these objects will be pathological even if the objects themselves are strictly, to the physiologist, external.

54. Testing my principle (that only endoparasites, not ectoparasites, are *per se* pathological) against the *International Classification* and *Stedman's Medical Dictionary,* I find only one misfit: lice. For mites, chiggers, fly larvae, sandfleas, leeches, tongue worms, and the rest, either the organism enters skin or other tissues, or the disease is defined as the bites or other injury, not the infestation. But

with head, body, and pubic lice the disease is defined by infestation. Since the eggs (nits) of scalp and pubic lice are firmly attached to our hair shafts, this seems to be a perfect borderline case of disease: organisms stuck to a dead part of the body surface. My precise ruling means that pediculosis and phthiriasis should be defined as the skin and systemic reactions, not the mere presence of lice and nits. And I am happy with this result, because one of the three listed diseases of lousiness is pediculosis corporis (or pediculosis vestimenti), which is defined as louse infestation of clothes. Obviously, clothes cannot be diseased. For lovely pictures of skin lesions from insect bites, scabies, and pediculosis, *see* du Vivier (1993), Chapter 14 ("Infestations of the Skin").

55. Of course, a competent hanging breaks the neck, which is a straightforward internal injury. But to give the example a run for its money, I will consider hanging as if it were garroting. I regret to have no information on the pathophysiology of tarring and feathering.

56. Hare (1986), p. 178. This ingenious argument may be philosophically unique in literally proceeding by hair-splitting.

57. Again du Vivier (1993) has a fine album of hair-loss pictures in Chapter 24, "Disorders of the Hair and Scalp." He shares Hare's view of disease, since he explicitly says that some types of hair loss or growth are diseases only by personal or social values. But in my opinion, despite the beauty of his atlas, du Vivier is abusing the medical vocabulary.

58. On fitness indicators *see* Williams (1978). Both the last two explanations apply to male hair, but perhaps the selection of genes for male scalp hair preserves female scalp hair as a byproduct.

59. In a spirit of charity, I have been ignoring the oddity of calling a practice a disease. On the BST a practice cannot be a disease; at best, it could only result from an internal

state that was disease, in this case presumably some kind of psychopathology. (A modern parallel might be smoking.) Historical analysis would show how far masturbation theorists distinguished between practice and internal state, disease and cause of disease. But insofar as they did not, they were wrong on that count as well, according to the BST.

60. (1975, p. 125). At the end of this essay, regarding drapetomania, Engelhardt says that "Cartwright's explanation failed because treating runaway slaves as free agents was a better account than treating them as subjects of fugue states" (138). But better how? If he means better cognitively, then he is in effect conceding that drapetomania was not, in fact, a disease, just as the BST implies.

61. On Lane *see* Comfort (1967), p. 125 ff. As in masturbation theory, for Lane "chronic intestinal stasis" caused a form of "autointoxication" called "alimentary toxemia," with such complications as "degeneration of heart, lungs and kidneys, uterine disorder, thyrotoxicosis, high blood pressure, insanity and tuberculosis" (pp. 126,127). Comfort estimates that Lane himself performed more than a thousand colectomies.

62. This is essentially the conclusion of the best discussion I know of masturbation mania, Comfort's chapter (1967, pp. 69–113). Comfort writes:

The gradual abatement of the outbreak had [by 1900] begun. It did not really die out completely until the 1940s with the statistical studies of Kinsey, and foci of infection still persist in old minds and books; 'excess', the remaining hook on which an anxiety could be hung, dies hard, but Kinsey's figures for the wide variation of sexual capacity in individuals, and the very high rate of orgasm in healthy young adolescents, remove the last rational point of adhesion. (110)

Already in the decades before the Kinsey report, both Robert Dickinson and Katherine Bement Davis had done statistical research on masturbation (Bullough 1994a,b, p. 309 ff). Indeed, when the Kinsey book appeared, the authors noted that while "many clinicians and educators" believed male masturbation universal, their own total figure was only 92% (Kinsey, 1948, pp. 498, 499).

Comfort adds that "social and cultural change" (111) also helped more liberal medical views of sex to take root, including psychoanalytic ones. No doubt social change facilitated the statistical inquiries that dealt the *coup de grâce* to 19th-century masturbation theories.

Since the first thing to find out about masturbation would surely be its prevalence, one might almost say that 19th-century masturbation theorists literally did not know the first thing about their subject. For this and other reasons, if Victorian masturbation theory is a model of good medical science, then I doubt a rational person could believe any scientific theory or visit any physician. If it is a model of bad medical science, how can we draw epistemological morals from it? But to distinguish between good science, bad science, and pseudo-science would not fit the genial epistemic nihilism of Engelhardt's essay. Instead of being a model of scientific method, 19th-century masturbation science is better compared to Renaissance witch manias (Comfort, 1967, p. 111).

63. A similar example is Pawelzik's proto-oncogenes (1990, p. 17). He notes that probably intracellular control mechanisms break down frequently, saddling the BST with constant undetectable outbreaks of cancer which then "heal" by themselves. To me this is just another case of a defense mechanism (immune surveillance), a function triggered by a dysfunction. The BST treats oncogenesis as pathological even if it has no clinical effects. Ruse also has a defense-mechanism case in his hypothesis (1981, p. 716) of homosexuality as a "cure," via kin selection, for low fertility.

64. One should note psychiatric parallels to environmental injury, and the uncertainty over their normality or abnormality. The best example is grief. To lose a parent, spouse, or child is a massive psychic wound requiring an extensive process of emotional healing. As in physical wound-healing, grieving is a normal response to a major injury, so its resemblance to psychopathology is not surprising. Freud (1917) used mourning to illuminate melancholia, but one can also do the reverse. To some degree, grief is a clinical depression provoked by external injury. (The injury is the perception of death, not death itself, since unknown deaths cannot cause grief while false reports of death can.)

Freud remarks that "it never occurs to us to regard [grief] as a pathological condition" (p. 243). Engel (1961) argues to the contrary that grief is a disease, but his view is not yet accepted by psychiatry. For the DSM, uncomplicated grief is not a disorder. Instead, clinicians commonly distinguish normal grief from various kinds of pathological grief, such as chronic, delayed, or inhibited (Stroebe et al., 1987, pp. 8–21; 1993, pp. 44–61). "Normal grief" seems to be considered normal for two main reasons: (1) the statistical normality in a human lifetime of loved ones' dying, and (2) the belief that human capacity for love entails vulnerability to loss (Reznek, 1987, p. 96). On the BST, by the wound analogy, reason (1) seems inadequate to keep grief from being pathological. But reason (2) might work, if it is true that no being could possibly love as we do, yet be able instantly to redirect its love when its current lover died. This interesting claim and many other issues require lengthy analysis, so I will take no position here on grief's normality.

65. This point is important in psychiatry. A child may fit his birth family poorly in many ways. For example, a child of normal intelligence born into an intellectual family may

be rejected by his parents, or develop various symptoms due to the stress of continually trying to meet their unusual expectations. It is crucial to explain to such parents that the child is normal; the fault lies with his environment. Any psychotherapist could give dozens of similar examples of normal children maladapted to their families. A similar point is the early observation by family therapists of how often families designate one member to be the "sick one," while most or all of the family pathology actually resides in the others.

66. In contexts where either fits, "may" conveys epistemic possibility, "might" metaphysical. Perhaps Hitler might have won the war by invading Britain in 1940 (metaphysical possibility). It is certainly wrong to say that he may have done so (we know he didn't).

67. Ruse (1981, p. 715) does not separate these two senses of "species design." As Nordenfelt notes (1987, p. 190n), Ruse therefore applies a non-BST concept of species design to heterosis, which he is discussing as a theory of homosexuality. I similarly object to his treatment of the parental manipulation theory (p. 717). By contrast, kin selection (pp. 715, 716) could be viewed as a type of reproduction and thus as normal by the BST. Although I know no medical basis for such a judgment, it seems a fairly natural extension of medicine's existing views on reproduction.

68. As Hare suggests, the same analysis should apply to his example of warfarin-resistant rats (1986, p. 177).

69. For another example of the dangers of confusing adaptation with health, recall Engelhardt's position that unwanted pregnancy is a disease and the extraordinary deference shown in *Roe v. Wade* (410 US 113 1973) to "medical" decisions. The typical abortion is a medical procedure only in that doctors do it. It is certainly not medical in being directed at health, since a fetus is not pathology and unhappiness is not ill health. But any justified deference

to medical judgment depends precisely on the usually uncontroversial importance of health. Since 1973, partly because of this opinion, at least 25 million human organisms have been exterminated. Of course, that may be just fine, e.g., if they were nonsentient and if there is nothing immoral about killing nonsentient organisms. The point is that a clear concept of health shows that calling abortion a medical decision does nothing to resolve its morality.

70. Margolis, 1976, p. 247; Engelhardt, 1986, p. 170; Miller Brown, 1985, p. 317; Pawelzik, 1990, p. 18. In a footnote (p. 31), Pawelzik mentions postmenopausal vaginal atrophy and suggests that the BST implies that for women after menopause, sex is an "unnatural act"! But that is to assume what I would never assume, that the only function of human sexuality is reproduction. Reznek's view (1987) is that what is medically normal depends on a normative decision about "priorities." We could regard aging as pathological if we pleased (p. 94), but since there is no present prospect of doing anything about it (p. 97), we give problems like cancer a greater priority. I do not quite grasp how this theory fits the many untreatable diseases medicine has recognized. In any case, I see no medical basis for Reznek's view that any universal trait of human beings, even the inability to fly, could be called pathological if we chose to prioritize its removal.

71. Note, however, that Reznek sometimes has in mind a population-control, not a reproductive, function for his aging and death examples (p. 129). Population-control functions are biologically dubious in involving group selection. Furthermore, on the BST, group-survival-related functions are irrelevant to health. Reznek uses such examples against the BST because he has previously accepted Wright's analysis of function over my own. But I am not responsible for health counterexamples deriving from Wright's function theory, since I argued at length

(WF) that it is fundamentally misguided and creates counterexamples in nonmedical contexts too.

72. Besides the female octopus, who, Reznek says (1987, p.
103), lives nine times longer without the optic gland that
causes her to guard her eggs, many other organisms
can enjoy longer life through sterility. Rose (1991, pp.
105,168) cites the diverse cases of salmon, soybeans, fruit
flies, and the marsupial mouse. For example, male marsupial mice make a reproductive effort so intense, with such
furious fighting and lengthy copulation by the victors,
that all of them are dead shortly after mating season,
either from combat injuries or simply from the hormone
storm. Likewise, in one famous study, human male
eunuchs lived nearly 14 years longer than their intact
counterparts (Hamilton and Mestler, 1969). It appears that
if one were to judge normality by life expectancy alone,
in most sexual species maleness would be a disease—in
fact, a serious one. Yet it seems incontestable that castra-
tion produces a pathological condition.

73. Surprisingly for so careful and sophisticated a writer,
Reznek seems to err in assuming a two-way conceptual
link between disease status and medical treatment. His
main analysis essentially ends in two ringing declarations:

> Judging that some condition is a disease commits one to
> stamping it out. And judging that a condition is not a
> disease commits one to preventing its medical treatment.
> (1987, p. 171)

The latter statement is plainly wrong, unless Reznek means
to pledge himself to the struggle against contraception,
cosmetic surgery, and obstetrical anesthesia. (Contraception by drugs or surgery is "medical treatment" by his p.
163.) The first statement seems wrong as well. For most
of the history of medicine, no effective therapy existed
even for major pathology. And today countless types of

minor pathology remain about which nothing can or need be done, except insofar as pathology is almost always bad and any bad thing should be "stamped out" if one can do so for free. The entailments Reznek assumes between disease status and practical medical treatment do not exist. Inconveniently for the BST, however, these errors are inessential to his position. Deleting all of his therapy clauses would simply give him a stronger theory.

74. Another, less lurid, reproductive-cost example is Engelhardt's hypothetical about IQs above 140 reducing fertility (1986, p. 170). (Perhaps we might imagine men exerting selection pressure for dumb blondes.) Conservatively assuming that high IQ still confers its usual survival advantages, we have a tradeoff between the two apical goals. Up to a point, I should think both groups, the dimmer more fertile and the brighter less fertile, would be within normal variation. But if IQ above 140 made reproduction impossible, it would seem pathological. Nonetheless, many of us would choose the brains over the kids. I expect Engelhardt's and Goosens' examples will strike a responsive chord in many women. Many a young mother, after a day spent with preschoolers, feels herself turning into a sluglike creature with half her original IQ.

    Ruse (1981, pp. 712, 713) mentions yet another conflict between the goals of survival and reproduction: Dörner's theory that male homosexuals live longer than male heterosexuals.

75. Of course, most infections involve local tissue dysfunction. Thus they must be considered pathological unless, like the normal flora of the colon, they have a physiological function.

76. I do find it unfair to blame doctors if sympathetic and respectful efforts to talk patients out of false theories or harmful therapies fail. In the high-blood case, if, despite her doctor's explanations and warnings, the woman persists

in taking salty folk remedies which worsen her hypertension, that does not seem to me her doctor's fault. Rather, where scientific physicians have expertise, it is ultimately patients' fault if they disregard it. To view the "patient as person" implies not only patient rights, but also patient responsibility.

References

Agich, G. J. (1983) Disease and value: a rejection of the value-neutrality thesis. *Theor. Med.* **4,** 27–41.

American Medical Association (1961) *Standard Nomenclature of Diseases and Operations,* 5th ed., Thompson, E. T. and Hayden, A. C., eds., McGraw-Hill, New York.

American Psychiatric Association (1980, 1987, 1994) *Diagnostic and Statistical Manual of Mental Disorders,* 3rd ed., Washington, DC: 1980 [DSM-III]; 3rd ed., rev. 1987 [DSM-III-R]; 4th ed., 1994 [DSM-IV].

Bechtel, W. (1985) In defense of a naturalistic concept of health, in *Biomedical Ethics Reviews: 1985,* Humber, J. and Almeder, R., eds., Humana, Clifton, NJ, pp. 131–170.

Bigelow, J. and Pargetter, R. (1987) Functions. *J. Philosophy* **84,** 181–196.

Boorse, C. (1975) [DI] On the distinction between disease and illness. *Philosophy and Public Affairs* **5,** 49–68.

——— (1976a) [MH] What a theory of mental health should be. *J. Theory Social Behaviour* **6,** 61–84.

——— (1976b) [WF] Wright on functions. *Philos. Rev.* **85,** 70–86.

——— (1977) [HTC] Health as a theoretical concept. *Philos. Sci.* **44,** 542–573.

——— (1987a) [CH] Concepts of health, in *Health Care Ethics: An Introduction,* VanDeVeer, D. and Regan, T., eds., Temple University Press, Philadelphia, pp. 359–393.

——— (1987b) [PMS] Premenstrual syndrome and criminal responsibility, in *Premenstrual Syndrome,* Ginsburg, B. E. and Carter, B. F., eds., Plenum, New York, pp. 81–124.

Brenner, C. (1974) *An Elementary Textbook of Psychoanalysis,* rev. ed., Anchor Books, Garden City, New York.

Brown, R. (1977) Physical health and mental illness. *Philosophy and Public Affairs* **7**, 17–38.

Brown, W. M. (1985) On defining 'disease'. *J. Med. Philos.* **10**, 311–328.

Bullough, V. (1994a) *Science in the Bedroom: A History of Sex Research,* Basic Books, New York.

———— (1994b) The development of sexology in the USA in the early twentieth century, in *Sexual Knowledge, Sexual Science,* Porter, R. and Teich, M., eds., Cambridge, New York, pp. 303–322.

Buss, D. M. (1994) *The Evolution of Desire: Strategies of Human Mating,* Basic Books, New York.

Cassell, E. (1982) The nature of suffering and the goals of medicine. *New Engl. J. Med.* **306**, 639–645.

Champlin, T. S. (1981) The reality of mental illness. *Philosophy* **56**, 467–487.

Clouser, K. D., Culver, C. M., and Gert, B. (1981) Malady: a new treatment of disease. *The Hastings Center Report* **11**, 29–37.

Comfort, A. (1967) *The Anxiety Makers,* Nelson, London.

Culver, C. M. and Gert, B. (1982) *Philosophy in Medicine: Conceptual and Ethical Issues in Medicine and Psychiatry,* Oxford, New York.

Daniels, N. (1981) Health care needs and distributive justice. *Philosophy and Public Affairs* **10**, 146–179.

———— (1985) *Just Health Care,* Cambridge, New York.

Downie, R. and Telfer, E. (1980) *Caring and Curing,* Methuen, London.

du Vivier, A. (1993) *Atlas of Clinical Dermatology,* 2nd ed., Gower Medical Publishing, London.

Dubos, R. (1959) *Mirage of Health,* Harper, New York.

Engel, G. (1961) Is grief a disease? *Psychosomatic Medicine* **23**, 18–22.

Engelhardt, H. T., Jr. (1974) The disease of masturbation: values and the concept of disease. *Bull. Hist. of Med.* **48**, 234–248.

(1975) The concepts of health and disease, in *Evaluation and Explanation in the Biomedical Sciences,* Engelhardt and S. F. Spicker, eds., Dordrecht, Reidel, 125–141.

(1976a) Human well-being and medicine: some basic value judgments in the biomedical sciences, in *Science, Ethics and Medicine,* Engelhardt, H. T., Jr. and Callahan, D., eds., Hastings Center, Hastings-on-Hudson, New York, pp. 120–139.

(1976b) Ideology and etiology. *J. Med. Philos.* **1**, 256–268.

———— (1980) Doctoring the disease, treating the complaint, helping the patient: some of the works of Hygeia and Panacea, in *Knowing and Valuing: The Search for Common Roots,* Engelhardt, H. T. and Callahan, D., eds., Hastings Center, Hastings-on-Hudson, New York, pp. 225–249.

———— (1984) Clinical problems and the concept of disease, in *Health, Disease, and Causal Explanations in Medicine,* Nordenfelt, L. and Lindahl, B. I. B., eds., Reidel, Dordrecht, pp. 27–41.

———— (1986) *The Foundations of Bioethics,* Oxford, New York.

Erde, E. L. (1979) Philosophical considerations regarding defining "health," "disease," etc., and their bearing on medical practice. *Ethics Sci. Med.* **6,** 31–48.

Farrell, B. A. (1979) Mental illness: a conceptual analysis. *Psychological Med.* **9,** 21–35.

Feinstein, A. R. (1967) *Clinical Judgment,* Williams and Wilkins, Baltimore, MD.

Fingarette, H. and Hasse, A. F. (1979) *Mental Disabilities and Criminal Responsibility,* University of California Press, Berkeley, CA.

Fodor, J. A. (1964) On knowing what we would say. *Philos. Rev.* **73,** 198–212.

Freud, S. (1917) Mourning and melancholia, in *Standard Edition of the Complete Psychological Works of Sigmund Freud,* vol. 15, 1957, Strachey, J., ed. and trans., Hogarth, London, pp. 243–258.

Fulford, K. W. M. (1989) *Moral Theory and Medical Practice,* Cambridge, New York.

Gert, B., Clouser, K. D., and Culver, C. M. (1986) Language and social goals. *J. Med. Philos.* **11,** 257–264.

Goodhart, C. B. (1960) The evolutionary significance of human hair patterns and skin coloring. *Adv. Sci.* **17,** 53–59.

Goosens, W. K. (1980) Values, health, and medicine. *Philos. Sci.* **47,** 100–115.

Guthrie, R. D. (1970) Evolution of human threat display organs, in *Evolutionary Biology,* vol. 4, Dobzhansky, Th., Hecht, M. K., and Steere, W. C., eds., Appleton-Century-Crofts, New York, pp. 257–302.

Hamilton, J. B. and Mestler, G. E. (1969) Mortality and survival: comparison of eunuchs with intact men and women in mentally retarded populations. *J. Gerontol.* **24,** 395–411.

Hare, R. M. (1986) Health. *J. Med. Ethics* **12**, 174–181.

Hesslow, G. (1993) Do we need a concept of disease? *Theor. Med.* **14**, 1–14.

Holzman, P. S. (1970) *Psychoanalysis and Psychopathology,* McGraw-Hill, New York.

Hull, D. L. (1965) The effect of essentialism on taxonomy: two thousand years of stasis. *Br. J. Philos. Sci.* **15**, 314–326; **16**, 1–18.

———— (1974) *The Philosophy of Biological Science*, Prentice-Hall, Englewood Cliffs, NJ.

International Statistical Classification of Diseases and Related Health Problems (1992) 10th rev. World Health Organization, Geneva.

Isselbacher, K. J., Braunwald, E., Wilson, J. D., Martin, J. B., Fauci, A. S., and Kasper, D. L., eds. (1994) *Harrison's Principles of Internal Medicine,* 13th ed., McGraw-Hill, New York.

Kass, L. (1975) Regarding the end of medicine and the pursuit of health. *The Public Interest* **40**, 11–42.

Kendell, R. E. (1975) The concept of disease and its implications for psychiatry. *Br. J. Psychiatry* **127**, 305–315.

Kinsey, A. C., Pomeroy, W. B., and Martin, C. E. (1948) *Sexual Behavior in the Human Male,* Saunders, Philadelphia.

Klein, D. F. (1978) A proposed definition of mental illness, in *Critical Issues in Psychiatric Diagnosis,* Spitzer, R. L. and Klein, D. F., eds., Raven, New York, pp. 41–71.

Kleinman, A. (1988) *The Illness Narratives: Suffering, Healing, and the Human Condition,* Basic Books, New York.

Ladd, J. (1982) The concepts of health and disease and their ethical implications, in *Value Conflicts in Health Care Delivery,* Gruzalski, B. and Nelson, C., eds., Ballinger, Cambridge, MA.

Lavin, M. (1985) Doctors, psychiatrists and disease. *Soc. Sci. Med.* **20**, 535–543.

Lewis, A. (1955) Health as a social concept. *Br. J. Soc.* **4**, 109–124.

Lewis, D. (1979) Scorekeeping in a language game. *J. Philos. Logic* **8**, 339–359.

Lewontin, R. (1970) The units of selection. *Annu. Rev. Ecol. Systematics* **1**, 1–16.

Margolis, J. (1976) The concept of disease. *J. Med. Philos.* **1**, 238–255.

Martin, M. (1985) Malady and menopause. *J. Med. Philos.* **10**, 329–337.

Mayr, E. (1976) Typological versus population thinking, in *Evolution and the Diversity of Life,* Harvard University Press, Cambridge, MA, pp. 26–29.

——— (1988) *Toward a New Philosophy of Biology.* Harvard University Press, Cambridge, MA.

McDiarmid, A., ed. (1969) *Diseases in Free-Living Wild Animals,* Academic, New York.

Merskey, H. (1986) Variable meanings for the definition of disease. *J. Med. Philos.* **11,** 215–232.

Mischel, T. (1977) The concept of mental health and disease: an analysis of the controversy between behavioral and psychodynamic approaches. *J. Med. Philos.* **2,** 197–219.

Munson, R. (1981) Why medicine cannot be a science. *J. Med. Philos.* **6,** 183–208.

Murphy, E. A. (1976) *The Logic of Medicine,* Johns Hopkins, Baltimore, MD.

Nesse, R. M. and Williams, G. C. (1995) *Why We Get Sick: The New Science of Darwinian Medicine*, Random House, New York.

Newman, R. W. (1970) Why man is such a sweaty and thirsty naked animal: a speculative review. *Hum. Biol.* **42,** 12–27.

Nordenfelt, L. (1987) *On the Nature of Health: An Action-Theoretic Approach,* Reidel, Dordrecht, The Netherlands.

——— (1993) On the relevance and importance of the notion of disease. *Theor. Med.* **14,** 15–26.

Pawelzik, M. (1990) Disease as functional disorder: a critique of C. Boorse's "objective" theory of disease [in German]. *Analyse & Kritik* **12,** 5–33.

Pickering, N. (1991) Review of Fulford, K. W. M., *Moral Theory and Medical Practice. Philos. Invest.* **14,** 179–183.

Reznek, L. (1987) *The Nature of Disease,* Routledge and Kegan Paul, London.

Richman, R. J. (1975) Review of Antony Flew, *Crime or Disease? Philos. Rev.* **84,** 425–429.

Rose, M. R. (1991) *Evolutionary Biology of Aging.* Oxford, UK.

Rosenberg, A. (1985) *The Structure of Biological Science,* Cambridge, New York.

Ross, A. (1979) Sygdomsbegrebet. *Bibliotek for Laeger* **171,** 111–129.

———— (1980) Det psykopathologiska sygdomsbegreb. *Bibliotek for Laeger.*

Ruse, M. (1981) Are homosexuals sick? in *Concepts of Health and Disease: Interdisciplinary Perspectives,* Engelhardt, Caplan, and McCartney, eds., pp. 623–724.

Scadding, J. B. (1988) Health and disease: what can medicine do for philosophy? *J. Med. Ethics* **14,** 118–124.

Schaffner, K. F. (1993) *Discovery and Explanation in Biology and Medicine.* University of Chicago Press, Chicago.

Sedgwick, P. (1973) Illness—mental and otherwise. *Hastings Center Studies* **1,** 19–40.

Sober, E. (1980) Evolution, population thinking, and essentialism. *Philos. Sci.* **47,** 350–383. Reprinted in Sober, *From a Biological Point of View,* Cambridge, New York, 1994.

———— (1984) *The Nature of Selection,* Massachussetts Institute of Technology Press, Cambridge, MA.

Sodeman, W. A. and Sodeman, W. A., Jr. (1974) *Pathologic Physiology: Mechanisms of Disease,* 5th ed., Saunders, Philadelphia.

Sommerhoff, G. (1950) *Analytical Biology,* Oxford University Press, London.

———— (1959) The abstract characteristics of living organisms, in *Systems Thinking,* Emery, F. E., ed., Penguin, Harmondsworth, UK, pp. 147–202.

Spitzer, R. L. and Endicott, J. (1978) Medical and mental disorder: proposed definition and criteria, in *Critical Issues in Psychiatric Diagnosis,* Spitzer, R. L. and Klein, D. F., eds., Raven, New York, pp. 15–40.

Stedman's Medical Dictionary, (1995) 26th ed., Williams and Wilkins, Philadelphia.

Stroebe, M. S., Stroebe, W., and Hansson, R. (1993) *Handbook of Bereavement: Theory, Research, and Intervention,* Cambridge, New York.

Stroebe, W. and Stroebe, M. S. (1987) *Bereavement and Health: The Psychological and Physical Consequences of Partner Loss,* Cambridge, UK.

Symons, D. (1979) *The Evolution of Human Sexuality,* Oxford, UK.

Szasz, T. S. (1974) *The Myth of Mental Illness,* rev. ed., Harper and Row, New York.

Toulmin, S. (1975) Concepts of function and mechanism in medicine and medical science, in *Evaluation and Explanation in the Biomedical Sciences,* Engelhardt, H. T., Jr., and Spicker, S. F., eds., Reidel, Dordrecht, The Netherlands.

van der Steen, W. J. and Thung, P. J. (1988) *Faces of Medicine: A Philosophical Study,* Kluwer, Dordrecht, The Netherlands.

Wakefield, J. C. (1992a) The concept of mental disorder: on the boundary between biological facts and social values. *Am. Psychol.* **47,** 373–388.

——— (1992b) Disorder as harmful dysfunction: a conceptual critique of DSM-III-R's definition of mental disorder. *Psychol. Rev.* **99,** 232–247.

——— (1993) Limits of operationalization: a critique of Spitzer and Endicott's (1978) proposed operational criteria for mental disorder. *J. Abnorm. Psychol.* **102,** 160–172.

Walton, J., Barondess, J. A., and Lock, S., eds. (1994) *The Oxford Medical Companion,* Oxford, UK.

Weatherall, D. J., Ledingham, J. G. G., and Warrell, D. A., eds. (1996) *Oxford Textbook of Medicine,* 3rd ed., Oxford, UK.

Whitbeck, C. (1978) Four basic concepts of medical science, in *PSA 1978,* Asquith, P. D. and Hacking, I., eds., Philosophy of Science Association, East Lansing, MI.

Williams, G. C. (1966) *Adaptation and Natural Selection,* Princeton University Press, Princeton, NJ.

——— (1992) *Natural Selection,* Oxford, UK.

Williams, M. B. (1978) Sexual selection, adaptation, and ornamental traits: the advantage of seeming fitter. *J. Theor. Biol.* **72,** 377–383.

Wright, L. (1973) Functions. *Philos. Rev.* **82,** 139–168.

Wulff, H. R., Pedersen, S. A., and Rosenberg, R. (1986) *Philosophy of Medicine: An Introduction,* Blackwell Scientific Publications, Oxford, UK.

Wyngaarden, J. B. and Smith, L. H., eds. (1985) *Cecil Textbook of Medicine,* 17th ed., Saunders, Philadelphia.

——— and Bennett, J. C., eds. (1992) *Cecil Textbook of Medicine,* 19th ed., Saunders, Philadelphia.

Introduction

Having dealt first with the very question of whether or not one should consider philosophically the problem of defining disease, and having decided in the affirmative, this article turns first to the main positions taken on the nature of disease, distinguishing between the naturalist *and the* normativist *positions. Next, in an attempt both to throw light on the specific issue itself, but also to reflect understanding back on the theoretical philosophical analysis of the first part, attention is turned to the question of human sexual orientation, and it is seen how the various models of disease categorize those with a nonstandard (heterosexual) orientation. In conclusion, some critical remarks are made about the social constructivist thesis, which sees all such concepts as disease and health as artifacts of society, generally introduced to oppress certain powerless minority groups.*

Defining Disease

The Question of Sexual Orientation

Michael Ruse

Concepts of "disease," "health," "sickness," "illness" are obviously central both to medical practice and to its theory.[1-3] Naturally, they raise questions of considerable philosophical interest and in this discussion I want to contribute to an ongoing debate by comparing thoughts on health and disease (and sickness generally) against recent thinking about human sexual orientation and its putative causes.[4-7] In so doing, I hope to throw light separately on the medical concepts and at the same time add to an ongoing discussion of considerable social importance. I shall begin at the more general level and then move to the more specific; but, first, as a kind of prolegomenon, I want to raise an issue that has recently gained significance: whether or not one can or should properly talk of matters to do with "disease" at all.

Should One Consider Disease Concepts?

For the moment, let us take "health" to be a matter of proper or ideal functioning as a human being and "disease" and ill-health generally to be a matter of failure to achieve such functioning. Later, we shall consider more specific definitions, but these will do for the moment. Let us grapple with the fact that there is a considerable and growing body of literature that suggests that it is inappropriate even to consider questions about the nature of disease, with the implication or presupposition that perhaps we would be better if the whole topic were left undiscussed.[8–12] Various reasons are put forward for this position from which I shall highlight three. Briefly my position will be one of considerable sympathy with the intent of the critics but eventually, as will be seen, I shall argue that one may, and indeed should, properly and legitimately look philosophically at notions of disease and health and ill health in general.

The first critical argument is simply that the very idea of "disease" is in itself altogether too loose or flabby to warrant attention. The argument is that, particularly in this day and age, the notion of disease has been extended so far outward that it really makes little sense now to apply the concept at all. Perhaps (unlike the 19th century) we now no longer look on a propensity by adolescents to masturbate as a diseased form of desire, but we are hardly better with the application of the notion of disease to just about every social deviation or abnormality, whether it be car theft, incest, or susceptibility to overindulgence in food and drink.[6] Regretfully, argue the critics, when a term is used this loosely it becomes virtually if not literally meaningless, and hence had better be dropped entirely from the lexicon. Perhaps there was a time when notions like "disease" (and associated terms like "illness") were meaningful, but this is no longer the case.

The second argument agrees with the first, that often the notions like disease are applied indiscriminately and perhaps unwisely extended to new subsets of human activities or conditions. However, from seeing the notion of disease as becoming so loose as to have no bite at all, this criticism suggests that the concept has sufficient force that to label an individual as "diseased" is, in some very real sense, stigmatizing. Thus, for instance, if one speaks of a person who is somewhat on the short side as "diseased," then one is thereby labeling him or her as inadequate in some form or another.

Of course, relatedly, one is inviting all sorts of people to move in and to "remedy" the situation. As with the horrific medical practices invoked in the last century to curtail adolescent masturbation, so for instance today we find otherwise perfectly normal and healthy people subjected to all kinds of hormone and related therapy in the name of normality. Whereas previously a short person could have gone through life as satisfied with his or her height as any of us, now he or she is made to feel inadequate and is forced, either physically or at least socially, into doing something about it.

The third criticism is somewhat tangential to the first two, and could quite probably be combined with either. This is very much the criticism of the so-called "social constructivist," who argues that all human behavior, and indeed all human conditions, are in some sense an epiphenomenon of culture.[7-13] Such conditions have no objective reality in the sense of existing in the world independent of us. Hence, it is argued, since diseases clearly fall beneath the social constructivist net in some sense, one should recognize that a disease has no true existence. Thus, pretenses that diseases exist objectively are little more than gambits, usually gambits backed by power considerations in order to control or otherwise regulate the lives of fellow human beings. This being so, we would probably be better were we to recognize that disease has no real standing and forgo any attempts at proof of self-sufficient ontological standing.

As already noted, I have considerable sympathy with these arguments: Indeed, I would add to them myself. However, ultimately, I suspect that they are not well taken or at least that the counterarguments that can be brought to bear are at least as strong, suggesting that there is still a place and need for informed philosophical discussion of notions like "disease" and "health."

Start with the issue of the looseness or flabbiness of notions of "disease." Few can be insensitive to this charge. At this very moment outside the main psychiatric hospital in Toronto, the Clarke Institute, a large sign solicits funds, informing the passerby that one Canadian in four suffers from problems of mental health. If this were indeed so, then one gives way readily to visions of blood-lust-driven Canucks, raging across the snowy prairies, intent on dispatching the miserable seals of Newfoundland before they turn their frenzied natures on each other. However, nothing could be further from the truth. Not only are Canadians dull and humdrum people, but they rather pride themselves on this! It is simply not true that, by any reasonable reading of the notions of good or bad health, one in four of us is mentally deranged. The exaggeration, intended to illicit sympathy and (more importantly) donations, simply backfires.

However, having said this, one cannot deny that there are those among us who do seem afflicted in some way or another. One does not want to go to the other extreme on the sensitivity scale and suggest that no Canadian suffers from problems of mental health. Hence, even if one decides not to use terms like disease, one will have to use terms of a similar nature: "psychologically afflicted," "mentally troubled," or some such thing. Thus, since a rose by any other name will smell as sweet, there seems little reason to deny the use of the terms we have already mentioned, like disease and health. Because some people misuse the terms, for whatever motives, however honorable, it does not follow that all of us misuse the terms or indeed that we are forever destined to misuse the terms.

The second argument is an extremely powerful one. There is no doubt but that the use of the term "disease" or "sickness" can be very effective in the hands of those who wish to oppress or otherwise stigmatize others.[14] Certainly, if enough people tell an adolescent boy or girl that his or her methods of sexual relief are sick, one can virtually guarantee that by the time of adulthood a considerable load of guilt will have been installed. Nevertheless, let us not forget two things.

First, historically, it has not always been the case that the use of terms like "sicknesses" or "malady" have necessarily been used purely in the interests of oppression. Take for instance, the case of homosexuality in Britain in the 1950s. Liberal people of all kinds, straight and gay, spoke then of homosexuality as a sickness, but this was done very much in counter distinction to the prevailing belief that homosexuality was an act of will, and an evil one at that. In Britain in the 1950s, homosexuals (men) were sent to jail—rightly the laws were known as the "blackmailer's charter." By arguing that homosexuality was no free act of will but was part of one's inherent nature, however unfortunate, a powerful and ultimately successful movement was launched that brought about the reform of the antihomosexual laws in Britain.[15] Today, of course, we look back and (at best) smile and (at worst) weep for many of the beliefs that then passed as truisms, but the fact is nevertheless that the sickness or disease label proved very powerful in the removal of highly offensive and oppressive laws.

The second point to note is that, whether or not the philosopher out of some sense of moral purity decides not to use such notions as disease, these notions are still very much inherent in our society generally, and indeed their use in various ways is taken as a matter of social policy. This is not necessarily, I rush to add, because people want to control a group by labeling them as "diseased," but sometimes for the very opposite reason, i.e., they want to escape responsibility by denying the title of sick or disease to certain segments of society.

Once again referring to Canada, recently the Province of Ontario has elected an extremely right-wing government. One of the first acts of this government has been to redefine the notion of disease so that large segments of society that hitherto came under the heading of ill or otherwise afflicted can no longer qualify for the term. This means, as an immediate result, that such members of society are taken off the rolls, of the Unworthy of support because of "illness" category and put into the "unworthy of support because of slackness" category. Already the government is denying support to people simply on the grounds that they no longer qualify as sick, ill, or diseased. Whether or not this is a good thing (and in many cases it is hard to think of arguments that might be provided to suggest that it is a good thing) the fact is that people in power are using such notions for disease and defining redefining, and qualifying such notions as their own social and political ends. Hence, it is if anything immoral for those of us with philosophical training to refuse to enter the fray and add our expertise toward a resolution of the issues at stake.

Finally, with respect to social constructivism, let me again express considerable sympathy for the ideas and position. Nevertheless, as a sweeping indictment of the propriety of using terms like health and disease, it simply fails. Even if something is a social construction, indeed perhaps by virtue of the very fact that it is a social construction, this is not to deny that entity existence.[7] After all, the United States of America is in an important sense a social construction. We look on the United States as one entity, even though Hawaii and Alaska are places apart from the mainland. We do not consider, for instance, the present country of Canada plus the State of New York to be one entity, one country, even though in this case they do adjoin. In some sense, the United States is real in a way that Canada plus the State of New York is not. Yet the reason why we think that the one entity is real and the other is not is clearly not a matter of physical geography but rather a construction from societal attitudes and decisions past and present.

None of this is to deny that there may be a significant social factor in thoughts of disease and health, but it is to say that the social factor in itself does not deny real existence or the propriety of using and applying such terms. Hence, here, as before, there is no *prima facie* case against the use of such notions, and with this now established, I turn to philosophical thought on the notions.

Models of Disease and Wellbeing

Naturalism

There are essentially two approaches to questions of health and disease or sickness. One attempts to treat matters in what is initially a value-free sense, incorporating values only later. This I have myself labeled the *naturalist* approach, but one may alternatively refer to it as the *empiricist* approach or the *positivist* approach.[6,16,17] (I do not intend to suggest that all naturalists are necessarily positivists, but they do represent one end of the philosophical spectrum.) The other approach sees the values coming in right from the beginning and denies that it is possible so have some sort of value-free enterprise even initially. This approach I have elsewhere labeled the *normativist* approach. Recently, the two approaches have been labeled the "Cartesian" and "Hegelian" approaches, respectively, but whether these particular philosophers (and only these) deserve to be used as markers seems to me to be a matter of some historical question and need not concern us here.[18]

The naturalist approach, as I have said, attempts initially to approach matters in a nonvalue-laden fashion. In particular, the notion of disease, the concept of disease, is defined without respect to the implications for the bearer—whether they be good or bad, happiness-generating or otherwise, or anything else of this emotive nature. Essentially, a healthy state is taken to be one of proper functioning, that is to say, proper functioning for the species *Homo sapiens*. A diseased state is taken to be one that, in some sense, interferes with this proper functioning. It will be noted that, in accordance with common medical practice, this means that notions of disease will cover not only such things as measles, but also broken legs and the like, even though in everyday language we would probably not use the term disease for the latter.

Of course, there is the question of what constitutes "proper functioning," but for the naturalist, in this day and age of Darwinian evolution, proper functioning is taken as something involving survival and reproduction.[19,20] A healthy person, then, is one who survives and reproduces according to the species norm, whatever that may be, and a diseased person, or a person with a disease, has in some sense a condition that prevents this normal achievement in some way.

Christopher Boorse, one of today's most articulate spokespersons for the naturalist position, speaks of some sort of notion of normality or satisfaction of capacities proper to the human species:

> There is a definite standard of normality inherent in the structure and effective functioning of each species or organism . . . Human beings are to be considered normal if they possess the full number of . . . capacities natural to the human race, and if these . . . are so balanced and inter-related that they function together effectively and harmoniously. (Boorse 1977, 554, quoting King 1945, 494)[21,22]

For Boorse, being healthy, then, is satisfying these standards of proper species functioning:

> 'A *normal function* of a part or process within members of the reference class is a statistically typical contribution by it to their individual survival by reproduction.' To be diseased is to fall below one's species' standards, and diseases are what brings this about. (p. 205)

It is not therefore necessary to have 50 children to be healthy, nor is it necessary to live to be 120. If the norm for human males is (say) about 75, then as long as one lives to about that age one can be said to have been reasonably fit and healthy. If one has two or three children the same is likewise true (*see also* refs. 23–25).

A diseased person, or a person with a disease, is therefore someone who has some affliction that in some way would detract from or prevent normal functioning. Presumably in this day and age in the western world one would not think of someone in, say, middle age with a cold as being particularly handicapped from a point of view of survival and reproduction; but presumably the point is that inasmuch as one would want to say that if a cold were a disease and not merely an inconvenience that it is something that truly has the potential for curtailing survival or reproduction: "It is a disease, but not particularly a bad one as opposed as something like lung cancer."

Parenthetically, I should say that some thinkers seem to find even this explanation quite failing for minor ailments like colds, and hence want in some sense to deny the dichotomy of disease and health as opposites. Carolyn Whitbeck[26] for instance, argues that it is quite possible for someone to be perfectly healthy, even in some meaningful sense to be enjoying full health, and yet at some level to have a disease. I will not stop to discuss this point here, although I confess that sympathetic though I am to Whitbeck's point, I do find her solution somewhat counterintuitive. A person with a head cold is not perfectly healthy, and this holds true even if you think that the naturalist account cannot truly deal with the fact that head colds are not life-threatening ailments.

To return to the main discussion, how then does the naturalist like Boorse introduce value notions? It is true that we philosophers can readily run up examples in which one would positively welcome a disease. Suppose one is a schoolboy with a major exam around the corner for which one has not studied. A massive head cold would no doubt be extremely welcome at such a point! However, these are special examples. How does one express the fact that, generally speaking, one simply does not want to have a disease? Simply, Boorse, as with other naturalists, puts the emphasis on the idea of "illness" to denote something that is undesirable.

> A disease is an *illness* only if it is serious enough to be incapacitating, and it is
>
> (i) undesirable for its bearer
> (ii) a title to special treatment; and
> (iii) a valid excuse for normally criticizable behavior. (Boorse 1975, p. 61)[23]

For Boorse, therefore, the opposite of health is disease, but only certain diseases lead one to say that one is positively ill. To be frank, I am not quite sure how Boorse would deal with the schoolboy's head cold. Under normal circumstances it would be "undesirable for its bearer" and it is thankfully "a valid excuse for normally criticizable behavior." I suppose what one would have to say an illness is something that under normal circumstances would be undesirable. Obviously, if the schoolboy knew that the next day there was a class trip to the circus, he would not welcome the head cold.

Normativism

Contrasting the naturalist, we have the normativist's position. This is represented in the literature by such people as Joseph Margolis[27,28] and the well-known philosopher of medicine, H. Tristram Engelhardt, Jr.[29] For these normativists, connection between disease and illness is not that between value-free and value-laden, but rather between cause and effect. A disease is something that brings on or causes an illness and for either to be considered as such they must be things that one does not welcome. Health, therefore, is the straight opposite of both notions.

> A first approximation to the theory of medical norms—**let there be stressed** that it is only a first approximation—has it that the body is composed of certain structured systems each of which has an assignable range of normal functioning. Defect or disorder of such systems relative to such functioning constitutes a sufficient condition of disease; illness, then, is reflexively palpable disease . . . the norms of health and disease tend to correspond—often in a disputatious way—with putative norms of happiness and well-being . . . To the extent that this occurs, it becomes difficult to treat the norms of medicine as altogether independent of ideologies, prevailing in different societies. (Margolis 1976, p. 245)[30]

Likewise, we find Engelhardt writing that

We identify illnesses by virtue of our experience of them as physically or psychologically disagreeable, distasteful, unpleasant, deforming—by virtue of some form of suffering or pathos due to the malfunctioning of our bodies or our minds. We identify disease states as constellations of observables related through a disease explanation of a state of being ill. (Engelhardt 1976, 259)[29]

I should say that, whether or not they truly be associated with the names of Descartes and Hegel, I see the naturalist and normativist positions as being part of a general broader dichotomy in philosophy. Naturalism or empiricism is, of course, a long-standing philosophy about the nature of science. Likewise, particularly recently under the influence of Wittgenstein, and then thanks to later thinkers like Thomas Kuhn,[3] one has had a very strong revival of normativism with respect to the epistemology of science. One has in many fields, therefore, and not just the philosophy of medicine, the dichotomy between those who believe that at least initially one can approach nature without making any interpretive judgements, and those who believe that all approaches to nature and to science in particular require interpretation—usually interpretation of a value-laden nature. (There is even a related debate today about the nature of law, with division between positivists, who think that one can approach laws without interpretation, and normativists like Ronald Dworkin,[32] who argue that all legal understanding requires value interpretation.)

There are, of course, variations on these two positions, with some trying to take positions incorporating the strengths of both sides.[26,33-35] There are indeed those who would argue that there is a third tradition of a more constructivist nature that stands distinct from both naturalism and normativism. Recently, this has been associated with the philosopher Nietzsche.[18] However, although perhaps here one can more readily sympathize with the identification of Nietzsche with constructivism than of Descartes and Hegel with naturalism and normativism, respectively, I am far from convinced that this represents an independent third option.

It is probably true that the naturalist or empiricist would have little in common with the constructivist. For someone like Boorse, for instance, an illness that leads to sexual malfunctioning—say an advanced form of syphilis—would not only qualify as a disease, but as a disease that is in some objective ontological sense quite independent of what society thinks about it. Malfunctioning of one's genitals is there to be discovered and requires no construction, social or otherwise whatsoever.

However, matters seem a little more complex in the case of the normativist. The normativist—certainly Margolis and Engelhardt—rather puts the whole notion of what constitutes functioning much more on a societal or cultural level. Take, for instance, something like a self-inflicted mutilation—if one might so call it—like a vasectomy. For Boorse this is a straight disease, because it prevents the transfer of sperm from male to female. For Margolis and Engelhardt, since in no sense would this necessarily move one away from happiness—it might indeed even be conducive to happiness—there is no reason whatsoever to speak in terms of illness or disease at this point. They would both presumably speak perfectly comfortably, and indeed intentionally, of the vasectomized male as being a totally and completely healthy human being. (Much more so than the miserable wretch who is terrified of the surgeon's knife and yet equally terrified of impregnating his wife for the umpteenth time).

But surely one would want to argue that in such a case as this latter, the whole notion of health and disease is something very much dependent on a cultural perspective. In ancient Israel, to take an example, the very thought of a vasectomy would no doubt be taken to be an appalling sign of Jehovah's ill favor. Here, culture dictates no limit to the maximum number of children, no matter what one's already achieved number. In other words, *qua* normativist, the whole question of the judgment of vasectomy as a disease/illness or not is something that is entirely a function of our societal attitudes.

I am not saying that this is necessarily always the case. Presumably both ancient and modern societies would look on lung cancer as a disease, and one might want to say that there is some physical basis to this. Thus, certainly the normativist is not committed to saying that everything is social. Perhaps, then, the extreme social constructivist is truly a third option, but I do wonder how many social constructivists would be quite so pure in their philosophy as to claim that the disease status of lung cancer is entirely a matter of construction.

Parenthetically again, it is interesting that when constructivists are making their case, they invariably seize on such delightfully supportive examples as masturbation, which are clearly matters of societal attitude that stay well away from such issues as lung cancer.[36,37] I make this point without denying the strength of the constructivist position *per se*. For instance, something like epilepsy could well be an interesting border case for some societies. For instance, in Russia in the 19th century and earlier, epilepsy was considered very much more than a disease and indeed some mark of holiness in a way that it is certainly not considered in late 20th century North America.

Leaving constructivism on one side, we have now before us our two models of health and ill health. Before turning to sexual orientation, I want briefly to note that there are those who may well be happy with discussions of disease and health in the physical realm and prepared to allow the legitimacy of the applications of the concept, but who would draw above the line when it comes to mental health. In particular, some psychiatrists and others argue that one simply cannot argue analogically from, say, a physical ailment like measles to a mental ailment like schizophrenia. Indeed, they would claim that the very use of the notion of "mental ailment" is in itself contradiction. Ailments or diseases by their very nature can be applied only the physical realm. (The classic statement is Szasz.)[38]

I will not stay long on this point, simply because it has been answered well by others.[6,27,39] It is indeed true that going from the physical to the mental is an extension of the term. One is, if you like, arguing analogically or metaphorically. However, such extensions are far from being illicit. They are the very meat and drink of creative thought. We extend our concepts analogically or metaphorically all the time, and there seems little reason to deny the legitimacy of so doing at this point. To use an example taken by myself and others previously, we all know what Jesus Christ meant when he spoke of committing adultery in one's heart, even though clearly no physical sexual intercourse takes place. Likewise, there seems no reason, at least in principle, why one should not move from the notion of disease, whether defined naturalistically or normatively, in the physical realm, and apply it to the mental realm.

Certainly someone who has some sort of mental condition that, for instance, produces impotency and at the same time causes grave concern (as one is told was supposedly the case of the 19th century writer Thomas Carlyle) has some sort of disease/illness. This is just as much the case if he or she had been born with a genital malformation that both made sexual intercourse impossible and also made the bearer terribly unhappy. This is not to go to the other extreme, and to say that all things that have been called "mental illness" are in fact properly called "mental illness"—I have suggested above that this is far from the case—but it is to say there seems no reason why disease and illness terms should not apply to mental just as much as to the physical.

Sexual Orientation

I turn now to the issue of sexual orientation. As I have said earlier, I do so both to cast light on the issue itself and also to reflect light back on thinking about concepts of disease, whether naturalistic or normativist. As always in philosophical discussions, the first priority is that of definition and in particular that of deciding what one means when one speaks of "sexual orientation."

Following usual practice, I shall make a threefold division between one's *physical sex,* anatomical, chromosomal, or genetic, one's gender and in particular one's *gender identification,* and one's sexual orientation. By physical sex, I mean the matter of whether or not one is male or female: this matter in humans being associated with sex-specific genitalia, the nature of the chromosomes, and so forth. By gender and gender identification, I mean the sense one has of belonging to a particular sex, that is to say one's sense of maleness or femaleness. By sexual orientation, I mean the objects of one's sexual desires or lusts.

In particular, the person who is sexually directed toward members of the same physical sex as oneself is a "homosexual" and a person who is directed to persons of the opposite sex to oneself is "heterosexual." It is difficult to give an operational definition of what one means by sexual orientation, but as good as any other is that which refers to the objects of one's sexual fantasies, particularly when one is masturbating. (I introduce this notion to draw attention to the fact that it is quite possible for somebody to act in a heterosexual manner while being homosexually oriented, or conversely. It is well known, for instance, that people of a heterosexual orientation often switch to homosexual activity when incarcerated).

I shall take it without argument or of great interest here that problems with physical sexuality and with gender can not only lead to dysfunctioning but can lead to illness or ill health of a kind that we have been discussing in previous sections. Someone with ambiguous genitalia (a hermaphrodite) is probably going to be unhappy, to say nothing of the difficulties in achieving any species norms about reproduction.[41] Hence, on either naturalist or normativist criteria such people will generally qualify as not simply unhappy but sick in the broad sense of ill or diseased. (I am not saying that necessarily every hermaphrodite is unhappy, but assuming that most are, which certainly seems to be in accordance with the known evidence. Admittedly, it is notorious that medical people only see unhappy people. Happy hermaphrodites presumably do not arrive in the offices of sex counselors.)

Again, problems with gender identification almost certainly are going to be associated with unhappiness and with disease factors, however categorized.[42,43] Someone who feels himself to be a woman trapped in a man's body is almost certainly not going to be functioning particularly well in a sexual role, probably either heterosexual or homosexual. This is not to say that there are no exceptions: The well-known travel writer Jan Morris, who changed from man to woman, first fathered children. However, generally speaking, such people will not be great reproducers and again will not be very happy. Indeed, their unhappiness is such that they voluntarily succumb to the surgeon's knife in ways that makes most people feel quite uncomfortable.

I should add that the independent evidence is that such people—"transsexuals"—are generally not people of great mental stability or happiness and qualify as sick in some sense, however defined. It altogether escapes me how one otherwise sensible writer on these matters can claim: "transsexuals do not appear to be mentally ill. There is the world of difference between a man who claims he is a woman and one who insists he is Jesus Christ reincarnated: the latter betrays his insanity with every sentence he utters, while the former is so obviously sane that he can often persuade a surgeon to cut off his penis."[4] I am really not sure why it is any less crazy to want your penis cut off than to think you can guarantee the whole of (deserving) humankind eternal salvation.

This brings us now to the much more interesting (speaking from a theoretical viewpoint) question of sexual orientation. At once, I forestall obvious criticisms, namely that there is little need or purpose to anyone getting involved in such discussions, specifically about the health status of sexual orientation, however one might decide on issues of health status generally. Even if it can be agreed that one properly can talk in terms of "disease," say for physical ailments like measles and perhaps even for mental ailments like schizophrenia, it might be felt that there is simply no place for opening discussions about disease or illness in the case of sexual orientation. After all, has it not been decided—by votes of psychiatrists, no less[44]—that homosexuality is not a disease, and certainly no-one here is suggesting that heterosexuality might be a disease! (As a philosopher one has to put in qualifying comments, because surely someone suffering from compulsive Don Juanism might be extremely unhappy and certainly does not satisfy species norms with respect to reproduction. I will ignore these sorts of counterexamples and concentrate only on the norm.)

Let me say simply that I am not so sanguine as perhaps are others that the question has been decided definitively. I say this quite apart from confessing to a general philosophical aversion to having epistemological matters settled by democratic ballot. Whereas compromise might be needed in the real world, conceptual issues are not as readily or promptly decided by a simple vote. Of course, in this day and age, particularly at universities in North America, there is some political hostility to opening discussions of this nature, since they are felt to be invariably prejudicial. Yet, although one cannot deny that there are surely times when this has been so, again I can only say that politics is no way to make philosophical decisions.

More generally, whatever the psychiatric community may have voted to decide, and whatever certain pressure groups might wish one to decide, the simple fact of the matter is that among lay people, there is still considerable controversy about the health status of nonheterosexual orientation. Indeed, as I write this, there are reports from England (*Weekly Telegraph,* March 6, 1996) that the military will continue to bar homosexuals, on grounds (among others) of the threat of "contagion."

Hence, although recognizing that this is hardly a politically neutral discussion, no matter what one's conclusions may be, I shall now move forward to consider homosexual orientation as it engages with the models of health and disease given in earlier sections. It is true that there are those who would argue that equity demands that one should likewise consider heterosexual orientation. For the purposes of this discussion, nevertheless, I shall simply assume that this is not nearly as problematic.[45] This is not to deny that there are unhappy heterosexuals and that one's hetero-sexuality can itself be a function of disease and illness of some sort or another. Not only does one have Don Juanism, as already mentioned, but one has all the various kinds of sexual deviations, some of which surely qualify as mentally unhealthy in one sort or another—heterosexual necrophilia, for a start.

In order to begin this discussion, one needs to consider homosexual orientation at a number of levels: In particular, one needs a discussion at the phenomenal level and then at the causal level.[6] First of all, what do we know at a factual level about homosexuality that is pertinent to models of disease and health? Second, what putative causes have been offered for homosexual orientation, and how if at all do these causes interact with the above-given models of health and disease? For simplicity here I shall consider only two basic causal models for homosexual orientation. The first is Freudian psychoanalytic theory, together with certain variations, and the second is what I shall generally refer to as the biological approach—here, I have in mind par-ticularly a genetic approach, but I take it that such a genetic approach has to involve more than just genes and will in fact make its way felt (most particularly) through various hormonal levels affecting the human body at various stages of development or life.

Let us begin at the phenomenal level. What do we know about homosexual orientation, specifically that in the West? There is some debate about the proportion of people who have a homo-sexual orientation exclusively or nearly so, but the original Kinsey figures of about 10% for the human male seems fairly robust. Of these about half, or 5%, are exclusively homosexual and the other half, another 5%, are more or less exclusively homosexual.[40,46] One striking fact is that the figures for women from Kinsey on-ward generally come out significantly lower than those for males, although more recently there has been debate about this.[47] In particular, there are those who wonder about the extent to which these low figures were an artifact of society, with women having less say or control over their sexuality. Perhaps, therefore, we should expect to see the female figures more closely approximate the male figures in the future. (There does seem to be general agreement from Simone de Beauvoir[48] onward that female sexuality is somewhat more diffuse than male sexuality. Therefore, females both psychologically as well as physically can play roles as either heterosexuals or homosexuals.)

Considering those who are exclusively or nearly exclusively homosexual, the surveys that we have do suggest that they are much more likely to go unmarried, and certainly to parent fewer children. This is not to say there are no homosexuals with large families—there certainly are—but on average homosexuals do have far fewer children than heterosexuals, and many homosexuals never have heterosexual intercourse at all. Moving to more contentious areas, that of happiness and well-being, once again the figures allow for many exceptions, but generally speaking one cannot find a truly significant difference between the happiness levels of homosexuals, males or females, and those of heterosexuals.[40] There are certainly many unhappy homosexuals, but then again there are many unhappy heterosexuals. There are also those homosexuals who are unhappy with their sexuality in one sort or another, but of course there are also those heterosexuals unhappy with their sexuality in one sort or another.

It is indeed the case that one does find homosexuals who are unhappy because of their homosexuality, and one presumes that this is more marked than among heterosexuals who are unhappy with their heterosexuality. However—and note here the explicit appeal to society—there seems little doubt that much unhappiness with one's sexual orientation is a function of societal norms and demands. In a society in which homosexuality is freely practised and condoned there is going to be considerably less tension than in those societies in which it is illegal and punished. Moreover, the figures that we have from various surveys bear out this natural assumption. One finds that homosexuals seem to have a better self-image about themselves with respect to their sexuality in more recent years than they did in previous times, and in countries in which there is tolerance rather than otherwise.[50]

Of course, in this day and age one can hardly have any discussion of (male) homosexuality and happiness without mention of the matter of AIDS. However, without in any sense wishing to trivialize or underrate it, basically I shall set to one side a discussion of the factors it raises. Undoubtedly, however you define it, AIDS is a disease causing a terrible illness associated with great unhappiness, but these are matters not inherent in sexual orientation *per se*. Some heterosexuals have AIDS and many homosexuals do not have AIDS, and indeed practice sex in ways that they are unlikely ever to catch it. I suppose if one could show that, as part of their sexuality—perhaps seeking out thrills—certain people practice sex in such a way deliberately making them liable to infection from AIDS, one would need to introduce it as a integral factor into the discussion, but I shall ignore this far from pressing possibility here. AIDS as such tells us nothing about homosexual health or lack thereof.

Causal Models

We move now to putative causal models of homosexual orientation. We start with the classic Freudian explanation, which states that homosexuality is a function of inadequate or otherwise disturbed psychosexual development.[57] Supposedly, according to Freud and his direct followers, all humans are born bisexual and go through various sexual phases in the course of development. One begins orientated toward the mother and her source of nourishment, the breasts. Hence, one is in a heterosexual phase for the male, homosexual for the female. One then moves to a fecal or anal phase, where one is concentrating on one's own body: homosexual for both sexes. One then moves forward to some sort of fixation on one's genitals: again homosexual. Then, after many years of quietude, full mature heterosexuality emerges at adolescence.

If, however, for some reason one cannot move forward or one regresses, one may remain permanently in an earlier phase and one is thus, immaturely homosexual. According to Freud, reasons for this deviation are varied. They can be a simple function of role models. Freud suggested that the prevalence of homosexuality in ancient Greece was because upper-class young boys were raised exclusively in male societies by slaves. But, more commonly, particularly in modern society, Freud argued that repressive fathers and overdominant mothers cause homosexuality, particularly in males (*see also* Bieber et al. 1962).[52]

In particular, one is unable to resolve one's Oedipal strivings. Hence, being desperately in love with mother, and yet being barred from having full sexuality because of the incest taboo, instead of (as in normal heterosexuality) projecting one's desires onto other women, one regresses and projects one's desire onto a nonthreatening male. A homosexual (a similar kind of story is given for females) is not therefore abnormal, in the sense of doing something nonhuman. He or she is rather doing something that the psychologically mature person does not do. Although physically and socially mature, he or she is (at some sexual level) caught back or has regressed back to an earlier stage.

I should add that although Freud himself did not think of homosexuality as being unnatural, some of his followers have nevertheless argued precisely this.[53,54] They claim that bisexuality—the keystone to Freud's position—is in fact a pseudo-concept and that therefore all people are truly heterosexual. Hence, homosexuality involves some form of false consciousness. These critics argue, therefore, that the homosexual is not merely regressed to a juvenile position, but involved in some form of self-deception of one sort or another. One important implication of this revision is that therapy, at least in theory, can help turn some who are homosexually afflicted toward heterosexuality. This is in direct contradiction to Freud and his immediate followers, who were far from convinced that such a change would be possible. Indeed,

Freud himself argued that at most therapy could help one come to an acceptance of one's nature. It could not change it.[55]

Turning to the biological domain, something that has received much attention recently, the argument is that in some sense homosexuality is a function of one's organic nature.[4-6,56] Generally, the immediate cause of sexual orientation is thought to be atypical (from the species-norm viewpoint) levels of hormones, notably the hormone testosterone. It is now known fairly definitively that hormone levels have little or no affect on sexual orientation when one is an adult; but the key time is now pinpointed as the midpoint of fetal development (3–6 months). It is believed that the hormonal levels affect the developing hypothalamus, leading to adult sexual orientation. In particular, those boys who receive lower than normal levels of testosterone are more prone to homosexual orientation, and, conversely, those girls who receive higher levels of testosterone are more prone to homosexual orientation. (It is certainly not the case that someone who subscribes to such a biological thesis denies entirely the importance of social factors as well. Indeed, it may well be that one wants to combine both. The important thing is that one does allow for the significance of biological factors.)

A major question is why precisely one would have such changed hormonal levels. One suggestion is that these could be a function of external stress, for instance, as brought on by wartime conditions.[57] However, recently popular are suggestions that it is genetic differences that are significant in change in sexual orientation.[5,58,59] Indeed, it is now claimed that one can identify particular genes, the possession of which lead to homosexual orientation, and absence of which point one in the other direction. I should say that, as yet, these are highly controversial claims, as, indeed, are all the biological claims, specifically those about the effects of hormones on hypothalamus development.[60-62]

However, one can say that these suggestions are all part of vigorous ongoing research projects. Naturally, therefore, one is led into related biological questions about why particularly some people might have such genes and others might not have them. Several "sociobiological" hypotheses have been put forward.[16,17,63–65] Some suggest that the genes may be under tight control of natural selection. Others suggest that, although under the control of selection, the genes' homosexual-causing effects may be byproducts or deleterious and carried only because they are linked pleiotropically with other genes that do confer advantageous characteristics. Yet other hypotheses argue that, even if the genes are involved, they may have no adaptive or nonadaptive implications at all. They may exist, for instance, purely for reasons known by biologists as "genetic drift." That is to say that, basically, they just fluctuate in a population, more or less escaping the affects of natural selection.

I should say that, if discussion of the biology of the genetic aspects of homosexuality is controversial, those of the putative selective factors lying behind the genes are even more controversial.[66,67] It has been suggested, for instance, that perhaps homosexuals are akin to worker ants, which we now believe are maintained in populations because, in some sense, they aid more fertile siblings. (This is the mechanism of so-called "kin-selection.") On such a scenario, homosexuals are foregoing their own reproduction for the benefits of increased vicarious reproduction through close relatives—such relatives being aided in their reproduction by the homosexuals themselves. (For instance, by advancing the prospects of nephews and nieces.) Let me say simply that these are unproven hypotheses, more fun to play with, than having a known reflection in reality.[68]

There are many variations one can draw on these different causal models—Freudian and biological—and I do not intend to suggest that any one of them is definitive or that taking one position excludes another. However, sketchy although this discussion has been, it will enable us now to turn to the key question of how our various models of health and disease enable us to think about the status of homosexual orientation.

Homosexuality as an Illness

The reader will realize already that the particular philosophical stance one takes on the questions of health and disease are going to have serious implications for the conclusions one draws about the status of homosexuality. They surely lead, at least in part—I am inclined to say "in major part"—to the differences between peoples' value judgements about the orientation.[16]

Start with the phenomenal facts and run past them the two approaches, naturalistic and normative. From the naturalist point of view, there is no question but that taken as a whole, homosexuality has to be judged as a sickness or a disease, simply because one is going to have reduced reproduction. As I have pointed out, this is not the case for everyone, but on average one will find that having a homosexual orientation leads to fewer children.

On the other hand, it is not necessarily the case that the naturalist will judge the homosexual to be ill. If one is perfectly happy in one's situation, then there is no reason to think that one is any more infirm than someone who has had a vasectomy. Of course, the difference is that vasectomy is something that is freely chosen, whereas if we know anything at all of homosexuality, it is something pressed on you rather than made as result of a conscious decision, but from the point of view of happiness itself, there is no difference with respect to health status.

This is not to deny that perhaps some homosexuals will regret the lack of children, so it may well be that someone with a homosexual orientation will regret, perhaps bitterly, that he or she is not heterosexual. In such a case as this, it would surely be appropriate to speak of someone as having an illness—although presumably one might also think in terms of cure, by adopting children or (especially in the case of lesbians) by having them biologically outside the social frame of heterosexuality.

Turning now to the normativist position, it will be seen that there is absolutely no necessity to invoke the description of "disease" at all. One might well say simply that a person homosexually inclined has a particular orientation that is part of his or her nature, akin to being white rather than black, or male rather than female. In itself it is neither a handicap nor a boon, at least judged from a social point of view, because the normativist puts much greater emphasis on locating good functioning within the cultural realm, rather than exclusively within the biological realm, and inasmuch as a particular orientation is acceptable culturally, then there is little reason to make adverse health judgments at all. It is quite possible to say straight out that a homosexual is neither diseased nor ill, and in fact is perfectly healthy.

Of course this does not preclude the fact that on a normativist account it may well be that in a certain society, for whatever reason, being a homosexual will bring on great guilt and distress, in which case it is surely appropriate to speak of such people as being in some sense not just troubled, but even sick. I suspect that this is a conclusion that some might want to dispute. They would argue that, for instance, homosexuals in England in the 1950s were not sick at all. It was just that society regarded them as such. However, I am a little unwilling to rewrite history quite so dramatically. Indeed, guilt apart, if people were regarded as being sick, and indeed if they themselves thought of themselves as being sick, then I am sufficient of a social constructivist to want to argue that perhaps indeed they were sick. I am also sufficiently constructivist to say that changing this state of affairs does not necessarily involve changing physical facts, but rather societal attitudes, which is surely what has happened.*

Turn next to the Freudian situation. Let me remind you that Freud himself preferred not to think of homosexuals as being sick. He certainly categorized them as being immature; but in his opinion homosexuality was no disease; it was rather an age-inappropriate desire. It is true that for Freud one might be very unhappy with one's sexual orientation, in which case one might speak of an illness in this context, but he regarded the cure not to

*See Note on p. 166.

be that of changing the orientation, but rather in changing the attitude of the concerned individuals about their orientation. (Freud, of course, would have endorsed the comments made in the previous paragraph, that change might well involve societal change, as well as that of the individual.)

It is surely the case that—even if Freud's causal claims be well taken—the naturalist would disagree with Freud about the health status of homosexuality. One would have to judge homosexuality as a disease, since there clearly would be a falling away from societal reproductive norms. Whether the naturalist would have to judge the homosexual to be ill is, of course, a matter of the extent to which one thinks that the Freudian position is going to lead, or have a tendency to lead, to unhappiness. There is no reason why, in itself, someone who is immature should be unhappy.

On the other hand, given the fact that Freud suggests that a great deal of homosexuality is brought about by dysfunctional families—in the male case, dominant mothers and hostile fathers— one might suspect that often homosexuals will in fact be tense and unhappy about their general situation. Therefore, without wanting to say this is a general rule, my suspicion is that homosexuality, in the Freudian perspective given the naturalist viewpoint, might well incline to an illness. This is not to deny that, as Freud himself suggested, it is possible that one might be able to resolve one's tensions, not by changing the orientation, but by changing one's acceptance (or society's acceptance) of one's state, in which case one would achieve good health.

The normativist differs from the naturalist at this point. He or she would not necessarily want to have to speak of "disease" at all. Yet, there would be surely some overlap between the two philosophies, inasmuch as the normativist would have to take into account the extent to which a Freudian-type situation might be expected to lead to happiness. Once again, however, the possibility of cure in the Freudian sense seems perfectly open. (I stress that I am not suggesting that the Freudian necessarily thinks that the homosexual is ill, from either the naturalist or the normativist position. What I am suggesting is that, perhaps, given the putative causes of homosexuality, one might expect this to occur more commonly than in heterosexuals.)

Of course, in the case of the revised type of Freudianism, in which homosexuality is seen as veering from an innate hetero-sexuality rather than (as for Freud) a regression to an earlier bisexual phrase, then clearly one would expect to find that homo-sexuals are going to be tense and fraught about their situation. In this case, one would surely expect unhappiness, and so in either the naturalist or normativist position judgements of ill health would be appropriate. Expectedly, it is particularly among those who subscribe to this position that we find the most vocal advo-cates of "homosexuality as disease."[54]

It is appropriate at this point that I say categorically that I believe there are both conceptual and empirical problems with this revised Freudianism,[6] so I am not in any sense intending to suggest that this is a position that is well-taken. I introduce it into the discussion on the grounds that, before one tries to refute any position, one should first understand it. I would suggest that we are now in a far better position to see why precisely it is this segment of analysts that so violently argues that homosexuality is a deviation from full health. Given their position, it is virtually a necessary conclusion. We can see also what must be the next move of those of us who find this conclusion inherently trou-bling. We must look critically at the claim that homosexuality involves a betrayal of our universally shared heterosexual nature. If this fails—and I have argued elsewhere that it does[6]—then the revised Freudian's claims about the disease status of homosexu-ality also fails.

As it happens, many people, both those sympathetic to a biological approach and those critical, believe that this leads automatically to judgements of ill health, whatever one's philo-sophical stance on the disease/illness relationship (*see* especially ref. 5). As we can see now, this is surely a mistaken conclusion. There is no reason whatsoever to suggest that biology in itself will lead to a judgement of disease or ill health any more (or, of course, any less) than will a social or environmental line of causal approach.

It is true that, given a biological causal picture, the naturalist would probably make judgements of disease. This would be for the same reason as with a social causal picture, namely because one would expect that biology would normally reduce reproductive success (although *see* the following). Whether biology in itself would necessarily lead to unhappiness is an entirely different matter. Frankly, I see no reason whatsoever why, given a biological cause of sexual orientation, one should necessarily be unhappy, or happy, for that matter. Judgements of this nature depend very much on the way society regards and treats those who do not have heterosexual orientation. If society cherishes or regards well people of homosexual orientation, then they may well achieve perfectly happy and healthy status, even though their genes may be dictating their sexual interests entirely.

We must recognize how limited is our understanding of the ways in which genes do work in affecting human behavior. Nevertheless, if one takes seriously some of the suggestions of the sociobiologists, from the viewpoint of our discussion there are some very interesting possibilities. These arise particularly over the ways in which such apparently counter-reproductive characteristics as homosexuality might be promoted by natural selection. Remember the claim that, far from possession of homosexual genes being detrimental to one's reproductive prospects, they may in fact be highly advantageous.[17] Rather than reproducing directly, one might be reproducing vicariously through the agency of close kin (the already mentioned kin selection).

The idea here is that, as with worker ants, one is aiding one's gene sharers to reproduce more efficiently than they would otherwise, and so although one sacrifices one's own reproduction, in fact the proportion of one's genes in the next generation is higher than it might be indeed if one reproduced directly oneself. Obviously this is highly speculative in the case of homosexuality, and there is little if any direct evidence suggesting that homosexuals of either sex do indeed aid in the reproduction of close kin (although remember that this could be done indirectly, not simply through physical aids but through the opening of opportunities for siblings and others, as well as through the leaving of money and so forth).

But, assuming if only for the sake of argument that there is some truth in this mechanism, then from the naturalist point of view, if anything, homosexuality might well be an aid to health, rather than a sickness! This is admittedly a somewhat paradoxical conclusion, and I am happy to agree that at our present state of knowledge it is probably best regarded as "not proven." However, it does point us back to one very important general point about judgments regarding health and illness, whatever one's philosophical approach. Such judgements are not fixed once and for all, but are in major respect a function of our ongoing development of the relevant science, phenomenal and causal. As always where values are concerned, the final judgement is a function of the philosophy one is endorsing and the state of the world as one understands it by experience and theory.

Conclusion

I trust that this brief survey of the implications of different approaches to health and disease for human sexual orientation has convinced the reader that there are matters here of considerable theoretical interest, and moreover that by grasping the different approaches, one can throw considerable light on the different conclusions that people make about the health status of human sexual orientation. One's philosophical stance taken reflectively, or unreflectively, really does matter.

Also I trust that by now the reader can see that the whole question of the extent to which disease and health can be said to be or not to be functions of society and of the interpretations given by culture are far from simple matters.[7] It is clear that societal norms really do count. This is true for the naturalist, and even more for the normativist. On the other hand, we have seen little reason to suggest that a radical social constructivism is the uniquely proper approach to these matters. Whether or not one judges something like homosexuality to be sickness or disease or a healthy alternate way of life depends very crucially on societal attitudes and behavior.

However, societal attitudes are not the only factors involved here. We have seen no reason to think that people's different sexual inclinations are purely functions of society or of cultural attitudes. Even if you think this to be the case, no proof has been offered. Such an assumption is undoubtedly false if one takes a biological approach. But it is true also if one takes a social approach, such as that represented by Freudian psychoanalytical theory. The desires are real in themselves however caused, and they are real however one may regard them. Furthermore, for the Freudian of whatever stripe, one's underlying biological nature is a crucial factor in unraveling the full causal picture. As so often in life, social factors matter, but not exclusively.

Let me return to an original worry. Should one even consider questions of disease, illness, and health, or is the very inquiry in itself one necessarily giving way to the forces of prejudice and fear? I cannot deny that this may happen on occasion; however, my analysis of one phenomenon, namely sexual orientation, has surely shown that a deeper understanding of these concepts can lead to a more enlightened understanding of why people draw the judgements that they do. This surely is the first step toward tolerance of all in our society, whatever their nature. We have a powerful argument in favor of the case for looking at such concepts as disease and health, and an equally powerful argument against the case for ignoring or suppressing such discussion.

Note

I am a little worried about this constructivist claim. It is certainly possible for whole societies to be mistaken. The virtues of bloodletting are almost nonexistent, although for generations everyone believed in them. However, suppose you had a society with priests, thinking themselves as priests, and regarded as such, and yet in a new generation it was decided that these priests were improperly ordained and hence not real priests. I would continue to claim that they were priests in some real sense, and I would put sickness into this camp rather than that of bloodletting. All of this is regardless of the fact that sickness labels tend to be self-fulfilling, and people called "sick" feel that they really are sick.

References

[1]Caplan, A., Engelhardt, H. T., and McCartney, J., eds. (1981) *Concepts of Health and Disease,* Addison-Wesley, Reading, MA.

[2]Brown, W. M. (1985) On defining disease. *J. Med. Philos.* **10,** 311–28.

[3]Kovacs, J. (1989) Concepts of health and disease. *J. Med. Philos.* **14,** 261–277.

[4]LeVay, S. (1993) *The Sexual Brain,* MIT Press, Cambridge, MA.

[5]Hamer, D. and Copeland, P. (1994) *The Science of Desire: The Search for the Gay Gene and the Biology of Behavior,* Simon and Schuster, New York.

[6]——— (1988a) *Homosexuality: A Philosophical Inquiry,* Blackwell, Oxford.

[7]——— (1995) Sexual identity—reality or construction? *Indentity, The Herbert Spencer Lectures for 1992* (Harris, H., ed.), Oxford University Press, Oxford.

[8]Hesslow, G. (1993) Do we need a concept of disease? *Theor. Med.* **14,** 1–14.

[9]Kline, D. (1986) Health, disease and medicalization. *Int. J. Appl. Philos.* **3,** 85–88.

[10]Ladd, J. (1982) Concepts of health and disease and their ethical implications, in *Value Conflicts in Health Care Delivery* (Gruzalski, B., ed.), Ballinger, Cambridge, MA, pp. 21–40.

[11]——— (1986) Why call social problems diseases: a response to David Kline. *Int. J. Appl. Philos.* **3,** 89–92.

[12]Nordenfelt, L. (1993) On the relevance and importance of a notion of disease. *Theor. Med.* **14,** 15–26.

[13]Foucault, M. (1978) *History of Sexuality, I,* Pantheon, New York.

[14]Weeks, J. (1985) *Sexuality and its Discontents,* Routledge and Kegan Paul, London.

[15]West, D. (1966) *Homosexuality,* Penquin, Harmondsworth, Mddx.

[16]—— (1981a) Are Homosexuals sick? in *Concepts of Health and Disease* (Caplan, A. L., Engelhardt, H. T., and McCartney, J. J., eds.), Addison-Wesley, Reading, MA, pp. 693–724.

[17]—— (1981b) Are there gay genes? Sociobiology and homosexuality. *J. Homosexuality* **6(4),** 5–34.

[18]White, K. (1994) A sociological perspective on disease. *Explorations in Knowledge* **8,** 27–38.

[19]—— (1982) *Darwinism Defended: A Guide to the Evolution Controversies,* Addison-Wesley, Reading, MA.

[20]Maynard-Smith, J. (1978) The evolution of behavior. *Sci. Am.* **239(3),** 176–93.

[21]—— (1977) Health as a theoretical concept. *Philos. Sci.* **44,** 542–73.

[22]King, C. D. (1945) The meaning of normal. *Yale J. Biol. Med.* **17,** 493–501.

[23]Boorse, C. (1975) On the distinction between disease and illness. *Philosophy and Public Affairs* **5,** 49–68.

[24]—— (1976) What a theory of mental health should be. *J. Theory of Social Behavior* **6,** 61–84.

[25]—— (1987) Concepts of health, in *Health Care Ethics* (Van der Veer, D., ed.), Temple University Press, Philadelphia, pp. 359–393.

[26]Whitbeck, C. (1981) A theory of health, in *Concepts of Health and Disease* (Caplan, A. L., Engelhardt, H. T., and McCartney, J. J., eds.), Addison-Wesley, Reading, MA, pp. 611–626.

[27]—— (1980) The concept of mental illness: a philosophical examination, in *Mental Illness, Law and Public Policy* (Brody, B. A. and Engelhardt, H. T., eds.), Reidel, Dordrecht, The Netherlands, pp. 3–23.

[28]—— (1986) Thoughts on definitions of diseases. *J. Med. Philos.* **11,** 233–236.

[29]Engelhardt, H. T. (1976) Ideology and etiology. *J. Med. Philos.* **1,** 256–268.

[30]Margolis, J. (1976) The concept of disease. *J. Med. Philos.* **1,** 238–255.

[31]Kuhn, T. (1962) *The Structure of Scientific Revolutions,* University of Chicago Press, Chicago, IL.

[32]Dworkin, R. (1986) *Law's Empire,* Harvard University Press, Cambridge, MA.

[33]Merskey, H. (1986) Variable meanings for the definitions of disease. *J. Med. Philos.* **11,** 215–232.

[34]Porn, I. (1993) Health and adaptedness. *Theor. Med.* **14,** 295–303.

[35]Rizzi, D. and Pederson, S. (1992) Causality in medicine. *Theor. Med.* **13,** 233–254.

[36]Stein, E., ed. (1992) *Forms of Desire: Sexual Orientation and the Social Constructist Controversy,* Routledge, New York.

[37]Hacking, I. (1986) Making up people, in *Reconstructing Individualism: Autonomy Individuality, and the Self in Western Thought* (Heller, T., Sosna, M., and Wellbery, D., eds.), Standford University Press, Stanford, CA.

[38]Szasz, T. (1961) *The Myth of Mental Illness, Delta,* New York.

[39]Macklin, R. (1972) Mental health and mental illness: some problems of definition and concept formation. *Philos. Sci.* **39,** 341–365.

[40]Bell, A. and Weinberg, S. (1978) *Homosexualities—A Study of Diversity Among Men and Women,* Simon and Schuster, New York.

[41]Money, J. and Ehrhardt, A. (1972) *Man and Woman: Boy and Girl: The Differentiation and Dimorphism of Gender Identity from Conception to Maturity,* Johns Hopkins University Press, Baltimore, MD.

[42]Stoller, R. (1968) *Sex and Gender: On the Development of Masculinity and Femininity,* Science House, New York.

[43]Green, R. (1974) *Sexual Identity Conflict in Children and Adults,* Basic Books, New York.

[44]Bayer, R. (1981) *Homosexuality and American Psychiatry,* Basic Books, New York.

[45]Symons, D. (1979) *The Evolution of Human Sexuality,* Oxford University Press, New York.

[46]Kinsey, A. C., Pomeroy, W. B., and Martin, C. E. (1948) *Sexual Behavior in the Human Male,* W. B. Saunders, Philadelphia.

[47]Kinsey, A. C., Pomeroy, W. B., Martin, C. E., and Gebhard, P. H. (1953) *Sexual Behavior in the Human Female,* W. B. Saunders, Philadelphia.

[48]de Beauvoir, S. (1953) *The Second Sex,* Knopf, New York.

[49]Gonsiorek, J. (1977) Psychological adjustment and homosexuality. *J. Supplement Abstract Service* (Am. Psychological Assoc.) MS, 1478.

[50]Whitham, F. (1983) Culturally invariable properties of male homosexuality: tentative conclusions from cross-cultural research. *Arch. Sexual Behavior* **12,** 207–226.

[51]Freud, S. (1955) Three essays on the theory of sexuality, in *The Standard Edition of the Complete Psychological Works of Sigmund Freud,* vol. 7 (Strachey, J., ed.), Hogarth, London, pp. 125–243.

[52]Bieber, I., Dain, H. J., Dince, P. R., Drellich, P. R., Grand, M. G., Gundlach, H. G., Kremer, R. H., Rifkin, M. W., Wilbur, A. H., and Bieber, T. B. (1962) *Homosexuality: A Psychoanalytic Study of Male Homosexuals,* Basic Books, New York.

[53]Ovesey, L. (1954) The homosexual conflict: an adaptational analysis. *Psychiatry* **17,** 243–250.

[54]Socarides, C. (1968) *The Overt Homosexual,* Jason Aronson, New York.

[55]Jones, E. (1955) *The Life and Work of Sigmund Freud,* Basic Books, New York.

[56]Hamer, D. H., Hu, S., Magnuson, V. L., Hu N., and Pattatucci, A. M. L. (1993) A linkage between DNA markers on the X-chromosome and male sexual orientation. *Science* **261,** 321–337.

[57]Dorner, G. (1976) *Hormones and Brain Differentiation,* Elsevier, Amsterdam.

[58]Pillard, R. C. and Weinrich, J. D. (1986) Evidence of familial nature of male homosexuality. *Arch. Gen. Psych.* **43,** 808–812.

[59]Ferveur, J., Stortkuhl, K., Stocker, R., and Greenspan, R. (1995) Genetic feminization in brain structures and changed sexual orientation in male Drosophila. *Science* **267,** 902–905.

[60]Gershon, E. S. and Cloninger, C. R., eds. (1994) *Genetic Approaches to Mental Disorders,* American Psychiatric Press, Washington, DC.

[61]Greenspan, R. J. (1995) Understanding the genetic construction of behavior. *Scientific American* **4,** 72–78.

[62]Kupferman, I. (1992) Genetic determinants of behavior, in *Principles of Neural Science (3rd ed.)* (Kandel, E., Schwartz, J., and Jessell, T., eds.), Elsevier, New York, pp. 987–996.

[63]Wilson, E. O. (1975) *Sociobiology: The New Synthesis,* Harvard University Press, Cambridge, MA.

[64]—— (1978) *On Human Nature,* Cambridge University Press, Cambridge, MA.

[65]Weinrich, J. D. (1976) Human Reproductive Strategy. Ph.D. diss., Harvard University, unpublished.

[66]Kitcher, P. (1985) *Vaulting Ambition,* M.I.T. Press, Cambridge, MA.

[67]Fausto-Sterling, A. (1993) Genetics and male sexual orientation. *Science* **261,** 1257.

[68]—— (1988b) *Philosophy of Biology Today,* State University of New York Press, Albany, NY.

Introduction

The authors provide the following definition of a malady: individuals have a malady if and only if they have a condition, other than their rational beliefs or desires, such that they are incurring, or are at a significantly increased risk of incurring, a harm or evil (death, pain, disability, loss of freedom, or loss of pleasure) in the absence of a distinct sustaining cause.

This account is both precise and systematic, thus enabling a more fruitful discussion of controversial cases. It has some surprising conceptual connections to rationality. The harms (or evils) that rational persons would want to avoid are the very harms that, when they are sustained in a certain way, constitute maladies. So the concept of malady is grounded in universal features of rational human nature. Although this means that values remain at the core of the concept of malady, the values are not only specified, but they are objective and universal. The authors' account of malady includes mental as well as physical maladies. Their definition of malady not only includes all the harms that are collectively included in the disease-illness-injury terminology, but it also includes all of the other essential features that these terms share.

Malady

K. Danner Clouser,
Charles M. Culver, and Bernard Gert

Introduction

The concept of disease (which we subsume under our concept of malady) is an issue that has been weaving in and out of bioethics for a long time. There are some bioethical matters for which the concept of disease has seemed pivotal, and others in which it has seemed to play no role at all. Furthermore, not everyone agrees about when the concept is relevant and when it is not. But in any case, the concept of disease deserves an in depth explication since it is so central to medicine and as such can and does so easily surface in discussions in and around the health professions.

The concept of disease is interesting in itself, apart from any moral implications. It is the central concept of medicine and yet, at its core, it involves value issues, although what issues and to what extent constitute much of the debate about the concept. Compare medicine's basic concept of disease with the basic concepts of other sciences—cell, molecule, gene, neuron, electron, proton, positron—all of which are empirically, operationally, or contextually defined, without any element of values. Admittedly, medicine may not be a science, but it is certainly very closely related to science and is considered by many to be a science, so the possibility of a value element at its core is noteworthy.

Explication of the concept of disease (as malady) is important not only because it is so basic but also because it enters into bioethical issues. A direct and significant consequence of our analysis is that it makes clear that abnormality alone does not constitute a malady, and thus deviancy (for example, in sexual behavior) would not in and of itself be sufficient for a condition being a malady.

An Overview of Malady

"Disease," "illness," "sickness," "lesion", "disorder," and so on are often used interchangeably, but each has its own slightly different connotation. The "concept of disease" is generally meant to include all of them. Such all inclusiveness becomes awkward in certain contexts, especially in light of all the other conditions that "disease" must refer to: injury, wound, defect, affliction, syndrome, trauma, lesion, disfigurement. The specific terms referring to these conditions have connotations such that in a certain situation, one particular term may seem the most accurate label; for example, "wound,"if one were stabbed with a knife. It would be odd to refer to a leg broken in an accident as a "disease" and it would be equally odd to refer to chicken pox as an injury. Nausea or a headache would probably be called an illness rather than a disease, since "illness" focuses on symptoms and on the afflicted individual's awareness of pain and discomfort. On the other hand, "disease" suggests an identifiable underlying physiological process with an etiology and with distinct stages of development, all of which could be present without the victim even being aware of them. In fact, many people have diseases but are not (yet) feeling ill.

Additional confusion exists because the conditions to which these terms refer overlap to some extent and the terms do not have consistent connotations. Ordinarily syndromes advance to being diseases when the causes and stages are more clearly understood, yet often the syndrome label persists simply because of tradition, for example, Down Syndrome. An occasional pathology text-book uses the term "traumatic diseases," presumably to avoid the limiting connotations of each term, "trauma" and "disease." Another example of the arbitrary nature of disease labeling is the condition experienced by deep-sea divers who return from the depths too quickly. It is called either "caisson disease" or "decompression illness," yet essentially all the ill effects are caused by the cellular *injury* caused by nitrogen bubbles forming in various bodily tissues. Other serious conditions do not fit comfortably with any of the standard disease terms; for example, hernias, mental retardation, and allergies.

Because of the plethora of terms, with their overlapping meanings and special connotations, we propose to give a new sense to the relatively infrequently used general term "malady" so that it will include the referents of all of these other terms. As we use the term, "malady" refers to that which all the other disease words have in common: "injury," "illness," "sickness," "disease," "trauma," "wound," "disorder," "lesion," "allergy," "headache," "syndrome," and so on. What is it that they all have in common? Answering that question constitutes the work of this essay. It is, in effect, our explication of the concept of disease and, therefore, the rationale for the introduction of the term "malady." Our analysis of malady can be taken as an analysis of the expanded sense of "disease" as it is often used in medical text-books, namely to include the referents of all these other terms. We prefer to use the term "malady," because the expanded connotation for the word "disease" is a significant distortion of the ordinary use of that term among both health professionals and lay persons.

The concept of malady is not new; rather, our analysis of malady attempts to capture what is common in the meaning of all the disease terms by making explicit and thereby calling attention to the common features in the meaning of all these terms. We use "malady" to refer to this commonality, because it is an old but very useful term that carries its meaning on its sleeve—as will become apparent. There seems to be no word in the English language that serves this overarching purpose; yet there is a need for such a word in order to refer to all of these conditions without being locked in to one or another of their special connotations. The all-encompassing word "malady" also helps us stay focused on the search for what the individual terms all have in common.

During the last 25 years analyses of the concept of disease have dwelt on the term's purported subjectivity and value-ladenness as if those two terms necessarily went together. We make a strong case for the concept of disease being value-laden but nevertheless objective. Our goal is to put the concept on a more stable footing, less subject to whimsy and manipulation. In the past the disease label has been used to try to accomplish a variety of personal and political agendas; it has been a very malleable concept. It has been manipulated to force "treatment," to deny reimbursement, to block entry into our country, to forbid marriage, to restrict freedom, to enforce morals, and to "medicalize" a variety of human conditions.

An ordinary example would be alcoholism. When it is important to impress on the alcoholic that he must share some of the responsibility, that he must exercise some willpower, he may be told that alcoholism is not a disease, but more of a moral failing. On the other hand, when it is important for the alcoholic to be freed from a burden of guilt and blame it is emphasized that alcoholism is a disease, the connotation of which is that it is an unfortunate condition that has befallen the alcoholic, victimizing him, and requiring expert medical treatment. Our concept of malady makes it clear that alcoholism is a malady, even though the alcoholic must bear some responsibility for controlling the symptoms of that malady.

Our account eliminates as much of the subjectivity as possible, allowing much less room for manipulation. Our account of maladies is both precise and systematic, thus enabling a more fruitful discussion of controversial cases. Nevertheless, there are places in which some vagueness remains. Our account of those shows what causes the vagueness and why it is unavoidable. We also give an account of the "logic" of malady, which includes a systematic framework for its use and an explanation of the ambiguities in its use. Inevitably there will be borderline cases. Our explication shows what aspects of these conditions make them borderline and what about these conditions would have to change for them to be clear cases of maladies or clearly not maladies.

Our account of malady has some surprising conceptual connections to our account of morality. The harms (or evils) that rational persons would want to avoid are the very harms that, when caused in a certain way, constitute maladies. So, like morality, the concept of malady is grounded in universal features of human nature. Although this means that values remain at the core of the concept of malady, the values are not only specified, but they are also objective and universal. Our account of malady includes mental as well as physical maladies. Our definition of malady not only includes all the harms that are collectively included in the disease-illness-injury terminology, but it also includes all of the other essential features that these terms share.

Some Background

There have been many attempts to set out a formal definition of "disease," but among medical professionals "disease" has been used in a technical sense, not in its ordinary sense in which there is a distinction between a disease, an illness, and an injury. Consequently "disease" is often used interchangeably with "illness," and injuries are regarded merely as a subclass of diseases. What most formal definitions of disease have intended to define is what we call a malady, that is, they have taken "disease" to refer to injuries, illnesses, headaches, lesions, disorders, and so on.

A characteristic definition is the following one from a pathology textbook[1]

> Disease is any disturbance of the structure or function of the
> body or any of its parts; an imbalance between the individual
> and his environment; a lack of perfect health.

This offers three separate but presumably equivalent definitions.
According to the first definition, recently clipped nails and pub-
erty are diseases, as would be asymptomatic situs inversus (right-
left reversal of the position of some internal bodily organs). The
second definition is too vague to be of any use, and the third is
circular. Although the "disturbance of structure or function of the
body" is inadequate when considered as a complete definition, it
can be incorporated as a feature of a more complex and adequate
definition.

Another medical textbook definition states that

> … disease may be defined as deprivation or lack of ease, a
> discomfort or an annoyance, or a morbid condition of the
> body or of some organ or part thereof.[2]

Here two separate definitions are offered. The second one is
obviously circular and hence of no help. The first may characterize many illnesses but certainly includes far too much. It would
rightly include heartburn, an earache, and an infected toe, but it
would also include an overheated room, tight-fitting shoes, and
irritating neighbors. Our definition of malady avoids this problem by requiring as a necessary condition that the malady be a
"condition of the individual."

An early definition of disease expresses what many subsequent definitions have emphasized, portraying the disease as a
result of an unfortunate interaction of the individual with his
environment:

> Disease can only be that state of the organism that for the
> time being, at least, is fighting a losing game whether the
> battle be with temperature, water, microorganisms, disap-
> pointment or what not. In any instance, it may be visualized

as the reaction of the organism to some sort of energy impact, addition or deprivation.[3]

Thus, one wrestler held down by another or an individual getting increasingly annoyed by loud music would have a disease. Our definition of "malady" avoids the flaw in such definitions by requiring that the harm being suffered have no "distinct sustaining cause."

Several authors have also correctly identified various aspects of disease. Spitzer and Endicott include in their definition of "a medical disorder" that it is intrinsically associated with distress, disability, or certain types of disadvantage.[4] This definition is on the right track. It begins to sort out specific harms that are at the core of disease, although it fails to capture all of them.

Goodwin and Guze significantly improve on Spitzer and Endicott's list of harms in their 1979 definition and add a condition that most others fail to consider in their definitions. Goodwin and Guze define disease as

> ... any condition associated with discomfort, pain, disability, death, or an increased liability to these states. ...

The important added feature that will be discussed and clarified in our definition of malady is "increased liability," which is a necessary amendment if asymptomatic conditions, such as high blood pressure, are to count as diseases.[5]

The foregoing examples of definitions are typical of those found in medical and pathology textbooks. Our primary interest is not in criticizing them, but in using them to introduce our concept of malady; they show there is a need for a more adequate definition and they suggest and exemplify some of the flaws that need to be addressed.

One kind of definition in particular directly raises moral issues: the definition that makes abnormality a central feature of disease. Although an example of this kind of definition has not yet been cited, many formal definitions in the literature make this particular feature the central one in the concept of disease. For example, in the 1992 (3rd ed.) of *Medicine for the Practicing Physician*, Editor-in-Chief J. Willis Hurst, in his own chapter "Practicing Medicine," says, "A disease is defined as an abnormal process" and he defines abnormality as "a deviation from the normal range."[6] Although whatever concept of disease is adopted has wide-ranging repercussions for everything from establishing the goals of medicine to the distribution of health care in society, for us the "abnormality" definition of disease has the most immediate moral implications. As will be shown in our explication of malady, abnormality is neither a necessary nor a sufficient feature of disease or malady. Nevertheless, abnormality does play an important but limited role in defining some of the terms in our definition of malady, and consequently that role needs careful discussion and specification.

Constructing a Definition of Malady

Our plan for this section is to lead the reader through a series of thought processes in order to produce a definition of malady. This is in contrast to simply stating the definition and defending it against possible objections. We believe this "method of discovery" will help with comprehension of the problem, its subproblems, and their solutions. A step-by-step process should make the nuances and maneuvers more accessible to the reader.

Something Is Wrong

In the very broadest sense, when one of the disease terms correctly applies to an individual or, as we shall say from now on, when an individual has a malady, there is something wrong with that individual. As seen in our earlier discussion of textbook definitions, however, to note that "something is wrong" is much too inclusive. Many things can be wrong in an individual's life without his having a malady: for example, being in poverty, being neglected, or being in a runaway truck. But before specifying when having "something wrong" with oneself does constitute having a malady, what it is to have "something wrong" with oneself must be made clear.

What is wrong with someone who has a malady? A typical and frequent answer would be: he is in pain, he is disabled, or he is dying. But do these states or conditions have anything in common? What is the genus of which pain, disability, and death are species?

The answer is, they are all harms (or evils). Harm (or evil) is the genus of which pain, disability, and death are species. What characterizes these harms is the fact that no one wants them. In fact, everyone wants to avoid them, or at least all persons acting rationally want to avoid them unless they have an adequate reason not to. The definition of an irrational action is an action of one who does not avoid harm for himself even though he has no adequate reason for not avoiding that harm.[7] Although sometimes people choose to die rather than suffer unrelenting pain, or to endure great pain to avoid a certain disability, they are not acting irrationally because they have adequate reasons for choosing the particular harm (or evil) in question.

We use the term "pain" in a wide sense, so that it refers to the unpleasant feelings of anxiety, sadness, and displeasure, and all the other kinds of mental suffering. Similarly, disabilities are not limited to physical disabilities, but include mental disabilities and volitional disabilities as well. Examples of mental disabilities would be aphasia and dementia; volitional disabilities include compulsions and phobias.

Furthermore, death, pain, and disability are not the only basic harms. Two other significant harms are loss of freedom and loss of pleasure, so it should be anticipated that these also would play a role in maladies, even though they do not immediately suggest themselves when one considers maladies. Interestingly enough, there are instances in which one or the other of these harms underlie conditions that would intuitively be considered maladies. For example, individuals with an allergy can avoid the circumstances and the places that trigger the allergic reaction, so that in effect they appear not to be incurring any harm. However, their freedom with respect to those circumstances and places has been limited, so in fact they are incurring a harm, namely, the loss of freedom. The loss of pleasure, independent of the other harms, is fairly limited in scope, but an example would be anhedonia, which is the inability to feel pleasure. It is often associated with schizophrenia, although its presence is neither necessary nor sufficient to establish that diagnosis. However, any condition of an individual that was characterized solely by the loss of pleasure, even without the negative feelings of sadness or anxiety, would qualify as a malady. Examples would be a condition that led to an inability to experience sexual pleasure or a stroke that affected the limbic system, blocking out the ability to experience pleasure.

Thus far, then, a malady can be said to be a condition that involves the incurring of harms. All the harms are instances of the basic harms (or evils): death, pain, disability, loss of freedom, and loss of pleasure. These are harms that every person acting rationally wants to avoid. This explains why and in what way malady (or disease) is a normative term. The concept involves values, certainly, but they are objective and universal values. Like colors, these values are universal and objective because there is general agreement about them. Most people make a distinction between different colors, e.g., between red and green, in the same way; those who cannot are regarded as color blind. There is a similar universal agreement among people about the nature of the basic harms and about the undesirability, absent an adequate reason, of experiencing them.

Significantly Increased Risk of Incurring Harms

The notion of malady (or disease) involves more than a person currently experiencing a harm, e.g., pain or disability. Many maladies do not initially involve incurring a harm; they can be asymptomatic at first. Such a condition may be regarded as a malady, even though it is not yet causing any incurring of harms, if it is a condition that will definitely, or very likely, lead to the incurring of harms, e.g., a positive HIV status or high blood pressure.

There must be an identifiable condition of the individual that is associated with the predicted harm. It is not sufficient that the individual be at risk of harm simply by being in a statistical cohort of some sort. For example, if all families living within a 10 mile radius of an atomic power plant have a 0.05% risk of developing thyroid problems, it does not follow that all of these individuals have a malady. Similarly, women whose mothers or sisters have or have had breast cancer may be at increased risk of getting breast cancer themselves, but they do not all therefore have maladies. Unless the condition that leads to the increased risk is identified as a condition of the person, that person does not have a malady. Thus, our nascent definition of malady needs to include that the condition that involves incurring a harm or the significantly increased risk of incurring a harm is a condition of the person.

We use the adverb "significantly" to modify "increased risk" because it highlights not only that trivial increased risks are not sufficient to make a condition a malady, but also because it explains the source of some disagreement in labeling a condition a malady. People do not always agree on when an increased risk is "significant," and it is preferable that the locus of those differences be as obvious as possible. Other variables enter into determining whether or not to call something a malady on the basis of increased risk. A barely significant increase in the risk of incurring a serious harm would likely lead to labeling that condition of an individual as a malady, whereas a greater increase of a mild harm might not. For example, a condition of a 20-year-old associated with a 3–5% greater risk of dying before the age of 50 would probably be regarded as a malady, whereas a 3–5% greater risk of experiencing mild joint pain before the age of 50 probably would not. Still another variable could be whether or not there is a cure. If a cure is available, there is some advantage to labeling a condition as a malady, even if the increase in risk that justifies the labeling is for a relatively mild harm in the relatively distant future. Our point in stressing the role of "significantly" in the important component of our definition "significantly increased risk of harms," is to alert the reader to a source of reasonable variation in how and what conditions are labeled maladies.

A Condition of the Individual

"A condition involving the incurring of harms or the significantly increased risk of incurring harms" still includes too much. The condition must pertain only to what is within the integument of one's body. It is limited to what is contained within that zone marked by the outer surface of the skin and inward. This distinguishes the condition from the situation and circumstances the individual is in. For example, an individual could be in an elevator with a broken cable, hurling downward. That individual is at significantly increased risk of harm, yet it would be incorrect to describe him as having a malady, at least at the moment. Likewise, an individual who is in jail is incurring a harm, namely, a loss of freedom. Yet this surely is not a malady. It is a situation the individual is in, rather than a condition of the individual himself.

As was seen in the previous section discussing increased risk, it is important to emphasize that the condition that would eventually lead to harm has a locus, namely, in an individual. That is done in order to distinguish it from mere statistical possibilities, in which case there is no identifiable condition of the individual that causes the increased risk of incurring a harm.

In order to make sure that the malady is within the individual and not identified with some circumstance of the person, this aspect of malady must be more carefully described. Our construction of the malady definition started with talking about an individual concerning whom something was wrong. Occasionally we have talked about a "condition of an individual." It is tempting to speak of a "bodily condition" in order to emphasize that the malady must be in the body and not a situation the body is in. That phrase, however, would make it impossible to include mental disorders within the malady label because, although at least some mental maladies are very likely to be conditions of the body, they need not be considered to be so by everyone. Referring to "a condition of individuals" leaves its ontological status an open question but still individuates the locus of the malady.

On the other hand, "malady" not only has the same meaning in "mental maladies" as in "physical maladies," but it also has the same meaning when applied throughout the plant and animal world, at least insofar as these plants and animals are capable of incurring any of the basic harms. Therefore, the locus of maladies properly should be regarded as "a condition of individual organisms." That phrase, however, seems somewhat stilted and given that our primary concern is with human maladies, we shall continue to use the simpler phrase "a condition of individuals."

Except for Rational Beliefs and Desires

So far our developing but still incomplete definition of malady is "a condition of individuals such that they are incurring or are at significantly increased risk of incurring death, pain, disability, or loss of freedom or pleasure." But "condition of individuals" can cover a lot of territory, so a certain narrowing and certain exclusions will be necessary to trim the concept down to something more intuitively correct. Beliefs and desires are part of an individual's condition and certain beliefs and desires can have the effect of causing harm or increasing one's risk of incurring harms. A belief that one has lost all his money in the stock market will cause suffering, but that suffering would not be regarded as a malady. Similarly, desires to climb mountains, ride motorcycles, and fly gliders are conditions of individuals such that those individuals are at increased risk of incurring harms, yet having those desires would not be regarded as having maladies.

On the other hand, there are beliefs and desires that not only cause harms to be suffered, but are themselves symptoms of maladies, usually mental maladies; for example, irrational beliefs and desires. One might have a belief that he is being tortured by demons and thus he would be currently suffering, or a person might believe that he could fly, and thus he would be at increased risk of incurring harms. A desire to commit suicide to see what it would be like to be dead would put one at significantly increased risk of harm. These cases do seem like maladies, even though they involve beliefs and desires.

This is sorted out by making "rational beliefs and desires" an exception to the conditions of individuals that can be maladies. Our expanded definition would then read, "a condition of individuals, other than their rational beliefs and desires, such that they are incurring or are at significantly increased risk of incurring death, pain, disability, or loss of freedom or pleasure." As seen in the previous paragraph, irrational beliefs and desires must not be ruled out as being involved in maladies since they may, in fact, constitute a malady. A belief is irrational only if its falsity is obvious to almost everyone with similar knowledge and intelligence. A desire is irrational when it is a desire for any of the harms or a desire for something that one knows will result in incurring a harm to one's self, and the person has no adequate reason for that desire.

Distinct Sustaining Cause

Although the focus has been on clarifying the nature of the condition that constitutes the malady within the individual organism, it must be recognized that although having this condition is necessary for having a malady, it is not sufficient. There are many instances of conditions of individuals such that they are incurring harms, not because of a rational belief or desire, and yet no one would regard that condition as a malady. A wrestler could be experiencing the pain induced by an opponent's hammerlock; a gardener could be experiencing the discomfort of the relentless sun beating down; an individual might be trapped in a tightly closed space and therefore experiencing aching muscles and anxiety, yet none of these people may have a malady. Some kind of conceptual move is needed in order to distinguish these cases from those conditions that are regarded as maladies. Those harms being suffered by the individual because of factors within the individual need to be distinguished from those caused by factors outside the individual.

Many harms are, of course, caused by agents from outside the individual: for example, allergens, germs, car accidents, and sun rays. Nevertheless, as long as the external circumstances are perpetuating the harm, for example the anxiety is being suffered because one is in a runaway car, then that anxiety would not be regarded as constituting a malady. If someone suffers a loss of freedom because of something internal to the individual, for example because of allergies or a fear of heights or open spaces, then that person has a malady. But if freedom is restricted by being in jail, then he does not. When a change in the external circumstances would immediately, or almost immediately, remove the harm being suffered, the condition is not a malady.

To accomplish this conceptual move with a bit more finesse, the notion of a sustaining cause must be introduced. As long as the harms are sustained by a cause distinct from the individual, the harms do not constitute a malady even though they do involve a condition of that individual. But these causes must be clearly distinct from the individual in whom the harms are being experienced. Thus, malady includes as a necessary condition that the conditions of individuals giving rise to their incurring harms has no sustaining cause that is distinct from the individual.

A distinct sustaining cause is a cause whose effects come and go simultaneously, or nearly so, with the cause's respective presence or absence. A wrestler's hammerlock may be painful, but it is not a malady because when the hammerlock ceases, so does that individual's pain. Of course, if the pain persisted a considerable time after the absence of the hammerlock, then a malady would be present because the pain, initially caused by an external source, is now being generated by the condition of the individual himself. A malady is a condition of individuals such that, whatever its original cause, it is now part of the individual and cannot be removed simply by changing his or her physical or social environment.

Our definition of malady now looks like this: Individuals have a malady if and only if they have a condition, other than their rational beliefs or desires, such that they are incurring, or are at a significantly increased risk of incurring, a harm or evil (death, pain, disability, loss of freedom, or loss of pleasure) in the absence of a distinct sustaining cause.

The Role of Abnormality: Disabilities, Significantly Increased Risk, and Distinct Sustaining Cause

Thus far, maladies have been discussed independently of any reference to normality. It was originally explicated in terms of something being wrong with an individual. We have explained that all we mean by "having something wrong" is that one is incurring a harm or is at significantly increased risk of incurring a harm. We have also made clear what the condition of the individual incurring a harm must be like in order to count as a malady. Now, however, some aspects of the definition of malady make it necessary to refer to the notion of normality.

One reason for relegating this discussion to a separate section is to emphasize that, contrary to most definitions of disease, abnormality is, for us, neither necessary nor sufficient for applying the disease or malady label. Earlier in this article, we quoted a recent example of abnormality being presented as the essence of disease in a book on medicine for the practicing physician: "A disease is defined as an abnormal process."[8] This is both wrong and potentially dangerous. Making abnormality the essence of disease has serious moral consequences, which is why a clear account of abnormality's proper role is needed. Labeling sheer deviancy as a disease mislabels, misdirects, and eventually leads to mistreatment. However, normality is a necessary feature for determining what counts as a disability, what counts as a significantly increased risk, and even what counts as a distinct sustaining cause.

Disabilities and Inabilities

One of the basic harms that can constitute a malady is a disability. However, determining what counts as a disability requires use of the concept of normalcy.

How is it determined that someone has a disability? Most cases are clear: Unless it is a fetus, if the individual cannot see, hear, or taste, it has a disability. If an adult cannot walk or sleep, or has a limited range of motion in his joints, then he has a disability. The problems of labeling come in the borderline cases. Do individuals have a disability if they cannot run, or if they do not have a full and complete range of motion in all their joints, or if they have less than perfect vision? What if someone who is only four months old cannot walk? Or what if an individual cannot jump to the top of a three-story building at any age? What if an individual cannot walk more than a quarter mile without tiring? These are the kinds of labeling problems that need to be worked through in order to gain an adequate understanding of the variables at work in the use of malady labels.

It is clarifying to distinguish between the lack of abilities that are properly called inabilities and those that are called disabilities. That humans cannot fly is a clear example of an inability. No humans can fly. Further, there are some extraordinary abilities that very few humans have, but that does not mean that the rest of us, who do not have these abilities, are therefore disabled. It is at this point that one must look to what is the norm for the species. The lack of an ability to run a mile in four minutes is an inability rather than a disability. Even though there are a handful of humans who can actually run that fast, it is so far from the norm that there is no question that a person is not disabled because he or she cannot do it. Labeling is more difficult with more normal abilities. Just how far should one be able to walk or run and in how much time in order not to be considered disabled? Invariably one must consult the norm for the species as well as discount that which takes special training, for example, pole vaulting, jumping hurdles, or playing chess. Many athletes can do extraordinary things, but again that does not mean that the rest of us are disabled. For one thing, the athlete's feats would be outside the normal range of what people can do, and for another, they require special training that the rest of us may or may not have an opportunity, let alone a desire, to acquire. If an ability is present in only a small subset of the human species, then the lack of that ability is not regarded as a disability (and hence as a malady); and if an ability requires special training the lack of that ability is not thought to be a disability. In both these cases the lack of ability is more accurately described as an inability.

A baby is unable to walk when she is only four months old. Is this a disability? Many elderly individuals cannot walk even a hundred feet. Do they have a disability? The baby does not have a disability, but the elderly individual does. A clarifying conceptual move in this regard, and one that parallels intuitive understanding of the matter, is to conceive of a stage in normal human development when this or that ability is at its peak. Until she reaches that point, an individual may lack the ability but would not properly be said to have a disability. We would simply say that she had an inability to do such and such or the person could not yet do such and such: for example, the baby is not yet walking. If the person still does not have the ability in question after the time when it usually appears in the human species, or has had it but no longer does, then the lack of the ability would be regarded as a disability, and hence a malady. For common abilities, after particular points in human development, the ability in question is simply regarded as the norm. After that maturation point, whoever does not have the ability has a disability no matter how many of the population at that stage, for example, 95%, do not have the ability. It is not inconsistent to say, "It is normal for the very elderly to have a significant loss of flexibility." Just because it is normal does not mean that it is not a disability, at least not after one has reached that stage of life at which having the ability is the norm.

In summary, both inabilities and disabilities involve the lack of abilities. An inability is not a malady. A lack of ability is an inability if either the lack is characteristic of the species or of members of the species prior to a certain level of maturation, or the lack is caused by the lack of some specialized training not naturally provided to all or almost all members of the species. A lack of ability is a disability if one has reached a stage of life at which that ability is the norm for the species. Furthermore, having reached that point of maturation, one has a disability even if most others in that age group also have that same lack of ability. Abilities that only one gender has are not abilities that are the norm for the species, thus lacking that ability is not a disability in those of a different gender (men are not disabled because they cannot bear children), but if a male or a female lacks an ability that is the norm for males or females, then that person does have a disability.

Low Ability

That disabilities exist along a continuum sometimes leads to some vagueness in using the malady label. When should a low level of an ability be called a disability? The gradations are sometimes so gradual there is no definitive way to draw lines. Obviously the norms for the species must be relied on considerably, i.e., the norms apart from any special training to enhance abilities. For example, an individual who cannot walk even after the maturation point at which a member of the human species generally acquires that ability obviously has a malady. But how far must he be able to walk not to be considered as having a malady? If he is limited to a few steps, or even to 100 yards, he would be considered disabled. But a mile? Two miles? One must rely on a comparison to the ability of the vast majority of humans at their prime for that ability. If that ability is distributed along a normal curve, then those falling in that range, no matter how wide that range is, would be considered normal, that is, without a disability. Those few at the very narrow low-ability end of the curve would be disabled, although, of course, those few at the far end of the curve would not be. They would have super abilities. (This latter is another reason why abnormality should not be regarded as the essence of disease.)

Something of this sort is already done in general intelligence testing, where 100 denotes average intelligence. Those having somewhat lesser scores, i.e., 70–80, are not regarded as mentally disabled but merely as having lesser ability, but those scoring 69 and under are regarded as being mentally disabled to varying degrees. Everyone knows that this is somewhat arbitrary and that there is no bright line distinguishing those who have a lesser normal intelligence (for example, an IQ of 71) and those who are mentally retarded (for example, an IQ of 69).

An individual who had developed an extraordinary ability, e.g., the ability to run a marathon in less than three hours, and then subsequently lost that ability, would not thereby have a disability, even if she lost it because of disease or injury. As long as she still falls within the normal range of ability, she is not disabled even though her own ability is significantly less than it was. Only if she got to a point where she fell outside the normal range of ability, for example could barely walk at all, would she be said to have a disability, and that would be true no matter how many individuals her age have the same problem. The comparison group for making these determinations is always the human species at its prime for that particular ability.

Allergies, Abnormality, and Distinct Sustaining Causes

Many maladies become active only in the context of certain external circumstances. When an individual is suffering discomfort and pain during an allergy episode, there is no doubt about his having a malady. At first impression it might seem as though this is a case of a distinct sustaining cause, since if the allergen were removed, the bad effects of the malady would disappear. But usually that does not happen quickly, and the individual continues to suffer for a period of time even in a changed environment.

What about an individual who is in an environment free of allergens, does she have a malady? Certainly she has a condition that puts her at risk of a malady. If she is living in a place where there are no allergens, then she is not very much at risk, but she still has the malady. She is not only at some risk of incurring a harm, she is already experiencing a harm, namely, the loss of freedom. Her choices of environment are limited because of her condition. However much she loves living in Arizona and never wants to leave, her freedom has been limited by her condition. Hence she definitely has a malady, although obviously not a very disturbing one for her.

It is in this context that another need for the concept of normality is uncovered. It is necessary for determining which reactions to one's environment are normal and which ones are indicative of maladies. If all humans gasp for air in a densely smoked-filled room, then gasping for air in a densely smoked-filled room does not indicate that one has a malady. The harm is not regarded as being caused by something within the individual. If only a small number of individuals develop shortness of breath in the presence of a cat, then the problem is within those individuals. They have a malady; and the harm is regarded as being caused by their condition, not by a distinct sustaining cause. Similarly, if an individual becomes anxious in the presence of almost everyone she meets, then she has a malady. In certain circumstances almost everyone gets anxious, but if someone gets anxious in circumstances in which most others would not, then she has a condition that is causing the problem. Thus, abnormality is an important consideration here too in deciding whether someone has a malady, because it determines whether experiencing a particular harm does or does not have a distinct sustaining cause.

Significantly Increased Risk

As has been shown, "being at a significantly increased risk," is an important phrase in our definition and we want to forestall some conceptual confusions that may arise from its use. These possible confusions relate to the matter of normalcy in the species.

Earlier in the chapter it was shown why it was necessary to use the locution "being at significantly increased risk." Some conditions of the individual that need to be labeled maladies are not yet causing perceptible harm, but they nevertheless put one at significantly increased risk for eventually incurring harm. High blood pressure is a good example, as is presymptomatic arteriosclerotic disease. In these examples it is clear that the significantly increased risk is based on empirical studies showing that having the condition leads to incurring some harm in the future. In most of these cases even the causal mechanism is well understood. Thus, the clearest instances of "significantly increased risk" are empirically based.

However there is another type of situation commonly spoken of as "significantly increased risk." An example of this would be an individual who has characteristics that place him in a cohort group that is statistically at higher risk of having a malady at a later time. Perhaps a pattern of family members and relatives have developed the disease and he falls within this pattern or grouping of relevant characteristics. This also is an empirically based calculation, yet, because the causal mechanisms are not proven, the eventuality of the incurring of the malady is not as certain. Also, this kind of statistical calculation makes it premature to speak of a "condition of the individual" that puts him at increased risk. It is not known what that condition is, but only that this individual is related to others who eventually have developed the particular malady in question. All that is known is that there is a good possibility that he shares the condition, yet to be discovered and specified, that produces the malady. Does such an individual have a malady? He certainly would not be regarded as having a disease or an illness; at most one would say that he is statistically at risk for an illness.

Malady is used in a similar way; it would not be said of someone who belongs to a group that is at higher risk of incurring some harm, e.g., because he lives in a certain location, that he has a malady. On the other hand, if, as is often the case, it is known that he has a certain feature, e.g., high blood pressure, that in conjunction with other known or unknown factors, will produce harm, he would be said to have a malady. For example, he might have a gene that is known to be a necessary factor in causing a particular malady. The sufficient conditions to produce the malady may not yet be known, but if the individual, by virtue of having that gene, is at a significantly increased risk over what is normal for the species, then it would be said that he has a malady, although it is not clear that it would be said that he has the same malady as the one that he is at increased risk of incurring.

One of the possible interpretations of "increased risk" that it is important to avoid is the one that would apply to an individual who had been in extraordinarily good health and had now slipped from that peak condition. Perhaps she had been a highly trained athlete, who ate carefully and exercised vigorously, but now has ceased training and relaxed her nutritional regimen. She is now more at risk for a malady, e.g., arthritis, high blood pressure, obesity, than she had been before. We do not label this kind of increased risk as a malady. The increased risk criterion is not a matter of comparing two states of the same individual, but a matter of comparing the state in question with conditions known to put humans at increased risk.

The key to understanding the meaning of "increased risk" is seeing it as a comparison with what is normal for the human species. Given that the human condition itself puts one at risk for all kinds of harms, the only conditions that count as maladies are those that significantly increase the risk of such harms. Maladies designate those conditions of the individual that put him or her at significant risk of harm over and above what is normal for members of the species in their prime.

Society's Reaction

The final aspect of malady that employs the concept of normalcy is an unusual consideration. It might best be seen if stated as a problem: If a malady is a condition of the individual such that he suffers harms, then might not skin color be a malady? Skin color, after all, can lead to persons inflicting death, pain, disability, loss of freedom, and loss of pleasure on one another, but it is certainly wrong and counterintuitive to think of skin color, ethnic origin, or gender as a malady. On the other hand, how should a grotesque deformity, or a significant disfiguration caused by accidents or surgery, or an extreme abnormality of one's body, be regarded? Much of the suffering associated with these latter conditions also seems based on society's reaction to the condition, although here the malady label seems more appropriate. If the latter set of conditions are maladies, how are they distinguished from the first set, which are not regarded as maladies?

It may seem that in the case of skin color society's reaction could be regarded as a sustaining cause that is clearly distinct from the person, concluding that therefore skin color, gender, or ethnic origin is not a malady. But what of the deformities, disfigurations, and extreme abnormalities? Do these not also depend on society's reaction? These extreme deformities usually involve some pain and disability and that alone makes them maladies without even considering the way the deformity looks. For example, there can be organ and joint involvement in extreme abnormalities of size; there can be difficulties in using one's hands and legs in deformities; there can be disabilities of seeing, blinking, breathing, and hearing in disfigurations. So in general, the pain and disabilities connected with abnormalities and deformities would be sufficient to classify them as maladies.

Suppose, however, that there is no such disability or pain connected with a severe disfigurement. Would the disfigurement in itself still constitute a malady? If there is a "natural" or "universal" shock to others who first observe them (for example, a person without a nose), they do have a malady according to our definition. If this is not a learned, acculturated response, but rather is a basic emotional response of all humans on first encountering these abnormal features, then the condition is a malady. In this sense such conditions are similar to allergies. In allergies, a condition of the individual interacts with certain elements of the environment resulting in harmful effects on the individual. One can avoid that environment, but doing so would then constitute another harm the person would be incurring, namely, a loss of freedom. The natural, spontaneous reaction of other humans to these physical abnormalities is like the natural environment. The abnormalities of the individual, not the natural reaction of others, are regarded as the cause of the pain and suffering to the individual. As a general rule, any deformity, abnormality, or disfigurement that is highly unusual and normally provokes an unpleasant response by others should be regarded as a malady. That, of course, leaves a certain amount of vagueness, but the elements that are essential for deliberating about each case are known.

Skin color was dismissed as a malady because society's negative reaction was taken as a distinct sustaining cause, but it can now be seen that that reasoning is inadequate. Consider again the allergy analogy. The allergic reaction also has a distinct sustaining cause, but the condition of the individual is regarded as a malady because the overwhelming majority of the species do not suffer harm when they encounter that environment. So, are skin color and similar features, like ethnic origin, really maladies? They are not and cannot be because the cause of the harm is different from the disfigurement example in that social repugnance to different skin color is not normal for the species. Indeed in other societies, that skin color or ethnic origin might provoke a positive response. When the reaction is only characteristic of a society, not of the species as a whole, the reaction should be regarded as a distinct sustaining cause, and the cause of that reaction, the skin color, should not be regarded as a malady.

This implies that if animals responded viciously to a condition of certain individuals, that condition would be a malady since it puts the individuals at a significantly increased risk of harm and deprives them of freedom. We accept that implication. Some persons attract mosquitoes much more than others, and they would certainly look on that condition as a malady. All humans attract mosquitoes to some extent, so do they all have maladies? No, because attracting mosquitoes is normal and thus they do not have a condition that puts them at significantly increased risk of harm. This is an instance of our meaning by "significantly increased" risk, a risk that is "significantly more than is normal for the human species." This is how it is determined if it is the individual who has the malady or if it is simply an abnormal environment. For example, submerged in water or closeted in a smoke-filled room, all humans have difficulty breathing. Such a difficulty is caused not by an abnormal condition of the individuals involved, but by an abnormal environment. Our standard is what is normal for the species. It is the norm for the species to be unable to breathe under water or in dense smoke. Thus, the problem is not in individuals. However, if an individual cannot breathe because there is a cat in the room, then the problem is within that person, because it is normal for an individual to be able to breathe in such circumstances.

Thus, the norm for the species is essential for determining whether the lack of an ability counts as a disability or only an inability; it is also essential for determining what counts as a significantly increased risk of harm; finally, what is normal for the species sometimes even determines whether a harm has a distinct sustaining cause, and thus whether the person incurring that harm has a malady.

Some Special Concerns
Cross-Cultural Issues

One sometimes hears accounts of "diseases" in other cultures that appear to conflict with our own culture's conceptualization. For example, conditions that our society regards as maladies, other cultures might regard as a sign of beauty or a gift of the gods. St. Vitus Dance, a neurological disorder, has been claimed to be a visitation from the gods; dyschromia spiraccotosis, which our society regards as a bacterial infestation, others have regarded as a beauty mark. Is the labeling of maladies purely a culturally relative matter?

The relativity of maladies is analogous to the relativity of morality. Both maladies and morality are basically universal, because both involve those harms that all rational persons everywhere avoid unless they have an adequate reason not to. However, a different culture could have a rational belief that leads it to interpret a particular pain, disability, or loss of freedom differently from other cultures. Thus, another culture may welcome a condition that most cultures would regard as a malady, but this does not show that the underlying concept of malady is society-relative.

Individuals in our culture also, on occasion, welcome a malady: to avoid work, to avoid being drafted, to avoid being chosen for a dangerous mission, or to receive compensation, but the malady involved is still a malady if it, in fact, causes death, pain, disability, and/or loss of freedom or pleasure. St. Vitus dance would be a disability in any culture, yet it may still, on balance, be desired if it is thought to be a favor of the gods. In short, although maladies are intrinsically bad, they can be instrumentally good.

The matter can become even more complex. What are regarded as good or bad ends are sometimes a function of the beliefs of that culture. For example, some cultures have ceremoniously inflicted severe wounds on members of their group because of beliefs that these wounds are associated with significant goods: purification, the passage to manhood, or some other spiritual benefit. Although it seems doubtful that these wounds would be thought of as maladies by those cultures, one suspects that a closer look at behavior would reveal that those wounds were attended to in the same manner that similar wounds, incurred under different circumstances, would be managed. It also seems likely that a similar wound on some other part of the body–not involved in the belief system– would be regarded and treated as a malady.

There is another reason that the very same condition of an individual might not be labeled as a malady by all societies. Some societies may not know that a certain condition is a malady because it is so endemic in their culture that they believe it to be a normal feature of the species. However, a society can be mistaken about this, just as it can be mistaken about any matter of fact. For example, measles, since it was so endemic, was once althought by some to be simply a necessary stage of child development.

Distinct Sustaining Cause Within the Individual

Another point of ambiguity concerns distinct sustaining causes. The issue at hand is whether the harmful effects experienced by the person stop when the external cause ceases. The clear case would be the wrestler's hammerlock, or being in jail, for which the pain or the loss of freedom stops immediately and coincidentally with the cause being withdrawn. But sometimes there are lingering pains or disabilities. Here, as in many of these issues, our labeling is guided simply by practical considerations. Someone who coughs for one or two minutes, but no more, after leaving a smoke-filled room would not be said to have a malady. There would be little point in so labeling her momentary condition. If the coughing continued for a significant amount of time, even an hour, depending on its severity, it might be regarded as a malady. That would be because what had been a distinct sustaining cause is now seen to have caused a condition within the individual that continues independently of the original cause. Now the cause of the pain and suffering is not distinct from the individual. It would be similar to a wrestler's hammerlock if the pain continued for one or two days after the hold was released. Now the condition of the individual that is producing the pain is not distinct from that individual, but is a part of that individual. To state it somewhat more precisely, the individual has a malady if and only if the harm he or she is incurring does not have a sustaining cause that is distinct from the individual. That keeps us from having to locate it; rather it requires only that it not be in continuing dependence on the distinct sustaining cause.

Our account of a distinct sustaining cause may suggest that those causes are always external elements, that is, outside the person's body. However, an increased risk of harm may have a sustaining cause that, although "clearly distinct" from the individual, is nevertheless within that individual; for example, a cyanide capsule held in the mouth. What if the poison capsule is swallowed but is still undissolved? At what point does a "clearly distinct" sustaining cause become one not so clearly distinct from the person? A bit of heavily encapsulated cyanide inside a individual's mouth is a "clearly distinct" sustaining cause because it has not yet been biologically integrated into the individual's body and it can be easily and quickly removed. If it were removed the individual would instantaneously be rid of the risk of harm. If it is swallowed and in the body of the individual, so that it cannot be easily and quickly removed, it is no longer a distinct sustaining cause and if it causes a significant increase in the risk of harm, its presence is a malady.

Another example of a foreign substance in the body that is not a distinct sustaining cause is the defoliative poison dioxin. It may become absorbed in the body's fat tissue and not have a harmful effect until the individual loses weight and the dioxin is released into the body, becoming metabolized within the body's circulation. Thus, the individual whose body is storing dioxin certainly has a malady because he has a condition that increases the risk of incurring an evil. This case is even clearer than the encapsulated cyanide that has been swallowed, because the dioxin has been biologically integrated with his body, even though it has not yet caused any harm.

Biological integration of a substance means that the substance has become a part of the individual, unlike, say, a marble that a child has swallowed. Biological integration means that body cells are invaded and interacted with, biochemical exchanges take place, and/or body defenses react. However, biological integration is not necessary for something in the body to cease to be a distinct sustaining cause. A clamp or a sponge left in a body cavity during surgery may seem to be a distinct sustaining cause, because the substance has not become a part of the individual, but since these items will soon cause serious harms, removing them must be done quickly. Food stuck in the windpipe of an individual seems to be another example of a distinct sustaining cause. It must be removed with great speed because it can very quickly cause major harms.

The sponge in the stomach or the food in the windpipe as distinct sustaining causes may appear to be instances of internal distinct sustaining causes because they are not biologically integrated. For us, however, anything in the body that is difficult to remove without special training, skill, and/or technology does not count as a distinct sustaining cause. However, catheters and proctoscopes may be an exception to this, because they go inside the body and cause pain or at least increase the risk of pain. Nonetheless, even though they may be difficult to remove without special training, skill, and/or technology, they are not counted as distinct sustaining causes because they are part of a medical treatment, and if properly introduced and withdrawn do not cause an increased risk of harm. Naturally, if there is a residual effect causing harm, then a malady has been caused to that individual. If a chemical tracer or nuclear marker has been introduced into the individual's body for diagnostic purposes and if these agents put the individual at a significantly increased risk of harm, then the individual has a malady, a temporary mild iatrogenic malady. If the cause of harm or increased risk of harm is inside the body, and has (1) become biologically integrated or physiologically obstructive, or (2) cannot be quickly and easily removed without special training or skill or equipment, or (3) has not been purposely introduced and withdrawn as part of a medical treatment, then the cause is not a distinct sustaining cause and the individual has a malady.

It is clear that in persisting to the fine points of using the malady label, we are making distinctions at a level not ordinarily recognized or considered. The ability of our analysis of the concept of malady to provide guidance not provided by ordinary disease terminology is an indication of its clarity and precision.

Troublesome Borderline Cases

Throughout this article we have dealt with problem cases in the context of developing criteria for a malady. We have been able to show that some conditions should have a malady label, e.g., phobias and volitional disabilities, and others should not, e.g., those conditions that are merely unusual, such as transvestism. More rigorous criteria have been established to purge the labeling of whimsical or subjective influence. However, in this section we explicitly address some conditions that are difficult to classify. This discussion, while making additional clarifications, will function also as a working review, showing how the concept works in some problematic cases. Generally what will be seen in these borderline cases is that the labeling could reasonably go either way. There will be reasons to classify some of these conditions as maladies, and reasons not to do so. Although we present our present view on whether or not the malady label is appropriate for each of these conditions, we are aware that small and plausible changes in our definition would result in classifying these conditions in just the opposite way. For example, adding to our definition another exclusionary phrase, "excluding those conditions that are necessary for the continuation of the species," results in pregnancy not being classified as a malady.

Pregnancy

On our present definition, pregnancy is a malady since it is clearly a condition of an individual, other than her rational belief or desire, such that the individual is suffering, especially in the last several months, some pain and disability. Also, throughout pregnancy she is at a significantly increased risk of incurring these harms. Pregnancy does seem to pose a counterintuitive result to our definition of malady. Certainly pregnancy has not been regarded as a disease nor, for that matter, as a disorder, a trauma, or an injury. None of those terms seem quite right. On the other hand, illness, nausea, and "feeling terrible," are terms frequently associated with pregnancy, and medical insurance pays for their treatment.

Malady, and other disease terms, have a negative connotation that makes us hesitate to label pregnancy as a malady. Pregnancy has an intrinsic outcome that makes it misleading to say that it is a condition that people want to avoid unless they have an adequate reason. Since the adequate reason is built into the condition, it is quite plausible to say that pregnancy is an entirely normal condition that happens always to be accompanied by maladies. On the other hand, that it is clearly a condition of the individual that puts her at a significantly increased risk of harms provides a good reason to call pregnancy a malady. There are other linguistic considerations, however, that make it plausible not to regard pregnancy as a malady, e.g., the misleading negative connotations of the term. It is finally a matter of choice, and what choice one makes should not affect the rest of our analysis.

A bad reason for not labeling pregnancy as a "malady" or "disease" would be the claim that pregnancy is normal. Although pregnancy is certainly normal, abnormality is only mistakenly taken to be the essence of a disease or disorder, so one ought not conclude that since there is nothing abnormal about pregnancy, it therefore cannot be a disease or malady. Our emphasis in defining malady has been on the harms suffered as a result of the condition of the individual, and pregnancy seems clearly to be a condition of the individual that causes harm. Some might be tempted to regard the fetus as an internal distinct sustaining cause, in which case pregnancy, carrying the fetus to term, would not be a malady. Although eventually the fetus becomes distinct, during most of the pregnancy it is not only biologically integrated in the body of the pregnant woman, but it is not easily and quickly removable without special skill, technology, and training. Therefore, according to our criteria, the fetus cannot be an internal distinct sustaining cause, and so if one decided on purely theoretical grounds, pregnancy would be a malady.

Menopause

Menopause also meets the criteria of a malady, even though it might be welcomed by the individual. It is a condition of the individual that necessarily involves a disability, although the lack of the ability to become pregnant is often not unwanted. The fact that menopause is entirely normal at a certain age does not keep it from being a malady. Similarly, a vasectomy in a man would be regarded as causing a malady, an iatrogenic malady, even if the resulting disability of infertility is exactly what was wanted. Although it was decided by the patient and performed by a physician, the result is still a malady since it is the loss of an ability that is characteristic of the human species in its prime.

Menstruation

Menstruation is a more difficult determination, it being so entirely normal to the female of the human species. It has been claimed that prior to civilized society, it was not normally a painful condition. Even now it is not painful for many individuals. If a particular individual does not suffer, then for her menstruation is clearly not a malady, but to the extent that the condition does cause discomfort and pain, it seems as if it would be considered a malady. Inasmuch as menstruation is completely normal, it might be less misleading to say that it itself is not a malady, but that it is often accompanied by maladies. On the other hand, the more constant the connection between the occurrence of menstruation and the accompanying harms, the more apt people are to regard menstruation itself as a malady. The strength of the association between menstruation and incurring harms is an empirical matter that could play a role in whether the malady label is to be applied directly to menstruation or only to its side-effects.

Premenstrual syndrome (PMS) is clearly a malady, because its very meaning is that the woman is incurring harms. If she is not, then she does not have premenstrual syndrome. PMS is talked about in terms of the harms it causes, yet it is clear that not everyone experiences it. The harms can be both physical and mental, creating body pains as well as volitional disabilities and severe mood swings.

The phenomenon of teething would present a similar issue. Teething is a completely normal occurrence during maturation, yet it often causes a lot of pain, and in this case, pain is necessarily part of the condition. Teething is always accompanied by some discomfort and pain, and hence for that reason, in spite of its being completely normal, our definition would classify it as a malady.

There is a temptation not to call pregnancy, menopause, and menstruation maladies, not only because they are normal, but also and especially because some believe it would be insulting and degrading to women to do so.[9] Although there is something negative about labeling a condition as a malady, we do not believe that labeling pregnancy as a malady should lead pregnant women to be treated worse than if one did not regard them as having a malady. Indeed, since we emphasize that an individual who has a malady is incurring some harm, applying the term "malady" to a condition should properly lead people to show more empathy toward anyone with that condition.[10] Our account of malady does not support any discrimination against or valuing less anyone who has a malady. Picking and choosing where the malady label should be applied on the basis of such whimsical and subjective grounds may lead to more arbitrary and harmful applications of the term.

Shortness of Stature

Shortness of stature is similar to many contenders for the malady label, in that it would appear that the physical (or mental) characteristic in question is a disadvantage only in particular social milieus. In a society that values height, shortness is seen as putting one at a disadvantage. The disadvantage is most likely be stated in terms of loss of freedom: freedom to play sports, to become a model, or whatever. Certainly there is not necessarily any pain, disability, or increased risk of death. The only problem would seem to be falling short of a societal expectation or value by virtue of the condition of the individual. This provides no grounds for classifying such conditions as maladies.

If the shortness of height is very severe so that disabilities and painful conditions are present, then, of course, the condition would be labeled a malady. Earlier we dealt with issues that involve society's reaction to an individual's condition. In the case of shortness, any loss of freedom that is experienced is a direct result of a particular society's reaction, but unless there is a natural, universal human reaction of shock or revulsion to the condition in question, there is no malady. Everyone who is not a natural born athlete seems to be at a slight comparative disadvantage in our society. Yet surely these persons do not thereby have maladies. There is nothing in the condition of the individual that is wrong; the problem comes about only as that condition interacts with the values, beliefs, and structures within particular societies. That may be a social problem, but it is not a malady. Furthermore, whether or not the shortness of stature was caused by a deficiency of human growth hormone or simply by genetic inheritance makes no difference regarding whether or not it is a malady.

Old Age

On the surface it would seem that old age is the epitome of a malady: a condition of the individual such that he or she suffers pain, disability, loss of freedom, and pleasure or is at a significantly increased risk of all those harms, in addition to death.

However, on balance, we choose not to call old age a malady, even though it is certainly accompanied by many maladies. There are several reasons for not calling it a malady. It is not clear in what sense age is a condition of the individual. Cirrhosis of the liver, a broken leg, a colon polyp, a missing clotting factor, and teething are all conditions of the individual. Age alone, the mere passage of time, is not a condition so much as a fact about the organism, a fact like "the individual swam in the river yesterday," or "he has lived through two world wars." Cells of the body change with age and different organs of the body change in a variety of ways, although a lot depends on the environment that the cells and organs have individually lived through. Thus, it is not helpful to characterize an individual with respect to age alone. Individual systems, cardiovascular, renal, endocrine, and so on, may have deteriorated with age, but not necessarily, and, in any case, the harmful conditions are individually identifiable and on these grounds are considered maladies.

Artificial and Transplanted Body Parts

Although the matter of artificial and transplanted body parts is not a contentious issue in disease classification, it is nevertheless of some interest, if only to test the logic of our account of malady. Does the individual who has an artificial or transplanted body part still have a malady? Using the malady criteria in fact makes the labeling fairly straightforward. If an individual who had suffered from a chronic malady and, because of available therapy, is now no longer incurring or at increased risk of incurring any harm, then he no longer had a malady. There is nothing wrong with him. Of course, realistically, if the individual has received a transplanted organ, very likely he is at a significantly increased risk of harm, and hence would be said still to have a malady. Normally he would be at less risk of harm than he was before the transplantation, but the comparison for determining the relative risk is not with his previous condition but with the norm for the species. However, after the transplantation, this individual's malady would normally be less serious than his previous malady.

On the other hand, if the individual has an artificial replacement and now is not at a significantly increased risk of harm, then he has no malady. Nothing is wrong. For example, suppose a woman had hypothyroidism, and physicians were able to implant in her a lifetime supply of completely safe and effective replacement hormone, so that none of the harms of the hypothyroidism were present and there was no increased risk of harm. She no longer has a malady. The same would be true of a formerly diabetic patient who now has an indwelling insulin pump, providing that the harms of diabetes are gone and the risks of harm are no greater than normal for the species. An artificial hip and an artificial lens implant in the eye would be other examples.

Accordingly, if these artificial or transplanted parts developed problems, the individual would be said to have a malady. Now the individual would be incurring a harm or would be at a significantly increased risk of incurring a harm because of a condition of that individual and there is no distinct sustaining cause involved.

Advantages of the Concept of Malady

A subtle benefit of using "malady" in this new technical sense is that it is the first explicit term in any language with the appropriately high level of generality. No language that we have investigated (English, French, German, Russian, Chinese, or Hebrew) contains a clearly recognized genus term of which "disease" and "injury" are species terms. Each term in the usual cluster of disease terms has specific connotations that guide and significantly narrow its use. Of course, that is as it should be if specificity is desired and justified. "Disease," "injury," "illness," "dysfunction," and other such terms overlap somewhat, yet each has its own distinct connotations. The advantage of "malady" comes by way of the term's generality, by way of its including all those conditions whose terms have their own individual, although over-lapping, connotations. This by no means makes the old terms irrelevant; rather "malady" is useful precisely in those contexts in which generality is important.

One such situation would be when nothing is known about a patient's condition that justifies the connotations of any of the other terms (injury, disease, sickness, trauma, phobia, lesion, disorder, wound, and so on). All these words have connotations about the nature of the condition and how it was caused. Inappropriate use of these terms can lead to wrong expectations and hence lead one temporarily down the wrong path of diagnosis. An injury described as a disease, a trauma described as a lesion, a lesion described as a wound, or a sickness as a disease could all be misleading. "Malady" is general and non-committal with respect to these connotations. It is useful as a beginning point for labeling a phenomenon, without being diverted into one or another connotation. Thus, although it is not as informative as other disease words that indicate by connotation more about the condition's diagnosis, status, and cause, it is a useful term when none of these circumstances are known and one needs to avoid being locked into the presumptions entailed in the use of other terms.

The search for a general term initiated the question: What do all the human conditions designated by the various disease terms have in common? This is the question that this article has tried to answer, and in doing so, has arrived at an analysis of the variables that enter into the labeling of these various human conditions. Our account of malady led to the recognition that all maladies involve either incurring at least one of the harms, i.e., death, pain, disability, loss of freedom, or loss of pleasure, or being at a significantly increased risk of incurring them. Because these elements of malady are objective, that is, they would uniformly be avoided by all rational persons unless they had an adequate reason not to, the influence of ideologies, politics, and self-serving goals in manipulating malady labels is considerably diminished. The possibility for some subjectivity does remain, but given the explication of malady, what elements are open to subjective bias is able to be determined more precisely. In short, the definition makes clear what is causing any disagreement.

The concept of malady has values at its core, but the values are universal and objective. Thus, our explication shows the inadequacy both of regarding disease as being totally value free and of regarding it as heavily determined by subjective, cultural, and ideological factors.[11] We have also tried to show the logic of cultural influences in those few instances in which they do occur.

A major contribution of our account of malady is that it shows that abnormality is neither a necessary nor sufficient definition of disease or of related terms. Abnormality becomes important in certain contexts, namely, in the determination of disabilities, distinct sustaining cause, and increased risks, but it does not play the major role that many other definitions have assumed it plays. Labeling a behavior or a condition a malady simply because it is abnormal can lead to unfortunate consequences and is not in keeping with what is intuitively meant by any of the various disease terms, such as disorder or dysfunction. Deviancy is not sufficient for using any disease term and the consequences of so regarding it can be significant. One tendency in the medical-scientific world has been to establish a normal range for this or that (e.g., some component of the human body), and *ipso facto* to have "discovered" two new maladies—hyper- and hypo-this or that.[12] Our account of malady makes it clear that this use of abnormality represents a misunderstanding of the concept of disease.

A final advantage of our explication of malady is that its basic elements, concepts, principles, and arguments are the same when applied to mental maladies. The usual bifurcation between mental and physical maladies disappears. As has been seen in numerous examples throughout this article, significantly increased risk of death, pain, disability, loss of freedom or pleasure, and the absence of distinct sustaining causes are applicable to the mental domain as well as to the physical.

There are other areas of medical ethical concern that seem to revolve around a definition of disease. The concept of disease plays a central role in discussions about the goals of medicine; for example, when it is argued that the traditional goal of medicine is to cure or ameliorate disease. The point of that maneuver is usually to distinguish that view of a physician's role from the more modern notion of the physician as a body mechanic for hire.[13] The body mechanic is one who knows a lot about the workings of the human body and can be hired to modify, enhance, rearrange, or otherwise intervene to satisfy the desires of the patient (or, perhaps, "client" would be more appropriate). Instances of this would be artificial insemination and other more sophisticated reproductive technologies, genetic enhancement, steroids for building muscle mass, plastic surgery, cosmetic enhancements, and so on. Some have argued that the goals of medicine need to be reaffirmed using the concept of disease, since that was the traditional focus of medicine. The cure or amelioration of disease constituted the traditional duty of physicians, so how disease is defined becomes important to the argument.

For example, whether or not pregnancy is a disease might be important for determining whether an abortion is a proper job of a physician. Focusing on curing and ameliorating disease would free physicians from the many other roles they have acquired through the years: stress reduction specialist, marriage counselor, nutritionist, weight reduction advisor, and exercise and body building expert, but adding "preventing disease" to the job description of a physician might bring back all of these roles. Perhaps the concept of malady could contribute to sorting out the goals of medicine, but as indicated above, we think this is doubtful. Ultimately, no analysis of disease or malady will be determinative of what medicine chooses to provide, so we shall not pursue the argument.

A definition of malady could be important to insurance companies and HMOs in determining what services would be covered: breast reconstruction following cancer surgery, human growth hormone for shortness of stature, treatment for alcoholism, a fear of heights. Although certain services could arbitrarily be declared within or outside of coverage, a good understanding of the concept of disease could be a help in negotiating borderline cases and in developing precedent for what should and should not be included in coverage.

Inevitably a precise definition of disease is important in the matter of excuses. What conditions are appropriate for sick leave? What is it proper for a physician to excuse? What constitutes an acceptable legal excuse, perhaps for certain behaviors? When is workman's compensation appropriate? Most likely these matters will become significantly more complicated as the genetic basis of more and more diseases is discovered. Even in clear cases of disease, there may be confusion about when to say the individual has the disease, since it might be asymptomatic for many years. Then there will be problems because of the probabilistic nature of the causes of the disease, ranging from certainty to unlikely, and finally, ways will be discovered to enhance the individual through genetic intervention, so that there may be a strong tendency to regard the natural unenhanced properties as maladies. If height, intelligence, memory, or strength can be improved by genetic intervention, then those expressions of unenhanced height, intelligence, memory, and strength may come to be regarded as deficient or handicapped, which is to say, "diseased". The concept of malady would be a help in clarifying what variables were relevant to that determination.

A definition of disease is also counted on in the matter of just allocation of health care. Whether it is a matter of a right to health care or a right to equal opportunity or simply a societal decision to be charitable, all participants to this discussion seem to end up talking about a "floor level" of health care. This is the level of care to which everyone either has a right or a guarantee. How should that level of care be determined? Usually it is done by appeal to the definition of disease, that is, by that which has been the traditional role of the physician, and that, according to many, does not involve body enhancements, social engineering, or spiritual and emotional guidance. Thus, the definition of disease plays a significant role in discussions concerning the allocation of resources.

All the above instances of appeals to the concept of disease are really part of the general search for the goals of medicine. Whether or not one believes that search is worthwhile, our concept of malady might be helpful to the clarity of the project. The concept of malady includes only those conditions of the individual that are harmful, and this is just the kind of condition for which those discussing the goals of medicine have been searching.[14]

References

[1] Peery, T. M. and Miller, F. N. (1971), *Pathology,* 2nd ed., Little, Brown, Boston, p. 1.

[2] Talso, R. J. and Remenchik, A. P. (1968), *Internal Medicine,* C. V. Mosby, St. Louis.

[3] White, W. A. (1926) *The Meaning of Disease,* William and Wilkins, Baltimore.

[4]Spitzer, R. L. and Endicott, J. (1978) Medical and Mental Disorder: Proposed Definition and Criteria" in *Critical Issues in Psychiatric Diagnosis,* Spitzer, R. L. and Klein, D. F., eds., Raven, New York, pp. 15–39.

[5]Goodwin, D. W. and Guze, S. B. (1979) *Psychiatric Diagnosis*, 2nd ed., Oxford University Press, NY.

[6]Hurst, W. (1992) Practicing medicine, in *Medicine for the Practicing Physician,* 3rd ed., Hurst, W., ed., Butterworth-Heinemann, Boston, London, Oxford, p. 14.

[7]Gert, B. (1988) *MORALITY: A New Justification of the Moral Rules,* Oxford University Press, Ch. 2. *See also* Rationality, human nature, and lists, *Ethics*, vol. 100, no. 2, January 1990, pp. 279–300.

[8]Hurst, W. Practicing medicine, in *Medicine for the Practicing Physician,* 3rd ed., Hurst, W., ed., Butterworth-Heinemann, Boston, London, Oxford, p. 14.

[9]Martin, M. (1985) Malady and menopause. *J. Med. Philos.* **10,** 329–337.

[10]Gert, B., Clouser, K. D., and Culver, C. M. (1986) Language and social goals. *J. Med. Philos.* **11,** 257–264.

[11]For example, Sedgwick, P. (1973) Illness—mental or otherwise, in *Hastings Center Studies* **1,** 19–40.

[12]Bailey, A., Robinson, D., and Dawson, A. M. (1977) Does Gilbert's disease exist?" *Lancet* **1,** 931–933.

[13]Kass, L. R. (1975) Regarding the end of medicine and the pursuit of health, in *The Public Interest* (no. 40), pp. 11–42. *See also* Bayles, M. (1981) Physicians as body mechanics, in *Concepts of Health and Disease,* Caplan, Englehardt, and McCartney, eds., Addison-Wesley, Reading, MA pp. 665–675.

[14]The ideas expressed in this article are part of a work in progress by the authors. The work in its final form will be published as a book with the title *Bioethics: A Return to Fundamentals.*

Introduction

In this paper Agich defends a pragmatic theory of disease and argues that disease is best understood in terms of practical goals and results, rather than in terms of concepts or theories. In sketching what a pragmatic theory of disease involves, Agich first contrasts this approach with other accounts of disease language. Second, Agich argues that disease is essentially a practical concept that is related linguistically and socially to the language of illness and sickness. Understanding the general relationship between the language of illness, sickness, and disease helps us to appreciate the central pragmatic nature of disease concepts. Third, Agich discusses three versions of a common theme in recent treatments of disease, namely, that disease language should be reinterpreted in descriptive and nonevaluative terms. Agich argues that this treatment of disease language divorces it from the actual context of medical practice. Fourth, Agich discusses four general kinds of purposes that guide the use of disease language, namely, care, cure, control, and communication, and point out that these purposes are sufficiently plastic to embrace many of the philosophical concerns regarding the ontological and logical status of disease language. A pragmatic theory of disease incorporates the role of values in the definition of disease; to understand the role of values one must attend to the historical and social context within which disease language is developed and used. Agich concludes by arguing that a pragmatic theory of disease retains many common philosophical concerns with disease language, but locates them in a far richer and more detailed context. Such a theory in particular challenges philosophers to focus their unique skills in analyzing concepts and arguments on actual, complex nosological issues.

Toward a Pragmatic Theory of Disease

George J. Agich

In this paper I defend a pragmatic theory of disease and argue that disease is best understood in terms of practical goals and results, rather than in terms of concepts or theories. In sketching what a pragmatic theory of disease involves, I first contrast this approach with other accounts of disease language. Second, I argue that disease is essentially a practical concept that is related linguistically and socially to the language of illness and sickness. Understanding the general relationship between the language of illness, sickness, and disease helps us to appreciate the central pragmatic nature of disease concepts. Third, I discuss three versions of a common theme in recent treatments of disease, namely, that disease language should be reinterpreted in descriptive and nonevaluative terms. I argue that this treatment of disease language divorces it from the actual context of medical practice. Fourth, I discuss four general kinds of purposes that guide the use of disease language, namely, care, cure, control, and communication, and point out that these purposes are sufficiently plastic to embrace many of the philosophical concerns regarding the ontological and logical status of disease language. A pragmatic theory of disease incorporates the role of values in the definition

221

of disease; to understand the role of values one must attend to the historical and social context within which disease language is developed and used. I conclude by arguing that a pragmatic theory of disease retains many common philosophical concerns with disease language, but locates them in a far richer and more detailed context. Such a theory in particular challenges philosophers to focus their unique skills in analyzing concepts and arguments on actual, complex nosological issues.

Ontological, Theoretical, and Pragmatic Approaches to the Concept of Disease

The history of theories of disease is rife with attention not only to the question of the ontological nature of disease, but to questions of the explanatory character of disease concepts.[1] Recent philosophical treatments of concepts and theories have also tended to focus on ontological and explanatory issues involved in disease language and classifications of disease. Indeed, many contemporary philosophical treatments assume that disease concepts either refer to entities that are amenable to empirical, scientific description or are explanatory structures that serve a purely theoretical purpose within scientific medicine. Thus, a proper concept of disease would be a well-formed explanation that conforms to nomological ideals. Because most of these theories are constructed on a relatively abstract plane, they do not investigate whether specific concepts of disease actually meet their standards. As a result, many advocates of this point of view assume that only so-called medical diseases can truly be regarded as diseases, because they alone refer to physical processes that explain the phenomena of illness in scientifically adequate terms. Medical diseases are accepted as paradigm cases of disease, because they seem to exemplify the ontological element that is believed to be essential to disease language and to exemplify the ideal of scientific, nomological explanation. A concept of disease is thus regarded as philosophically adequate if it meets the explanatory ideals whether or not it succeeds in capturing the sense in which the concept is actually used in the practice of medicine.

The view I develop takes disease language as essentially connected to the care of sick individuals. Thus, disease language has to be understood as a product of practical, not theoretical reason. A pragmatic theory of disease is pragmatic in the quite literal sense that disease is concerned primarily with practical needs and results, rather than with the ontological or logical aspects of theories of disease. It is thus less concerned with ontological questions, at least in the way that they have commonly been represented in philosophical theories of disease. I thus generalize a point made by Georges Canguilhem, who pointed out that "the impetus behind every ontological theory of disease undoubtedly derives from therapeutic need."[2] This truism motivates the pragmatic analysis of disease language. Regarded pragmatically, the language of disease is seen as part and parcel of a broad historically and socially determined therapeutic response and approach to sick persons. The identification of disease states and the construction of disease concepts, accordingly, can be expected to reflect a broad spectrum of human interests that interplay in shaping the historical and social forms of response. Nonetheless, this diverse range of practical interests is ultimately related to four main kinds of pragmatic concerns associated with the practical needs of sick persons: care, cure, control, and communication. Thinking of disease in these terms is important philosophically, because it keeps our attention focused on the actual social practice of medicine within which concepts of disease function rather than on some idealized model of medicine or science. It also helps to focus philosophical analysis on the concrete problems and issues that are involved in the use of disease concepts in the practice of medicine.

In this paper I only argue that a pragmatic view of disease construction and classification is in general plausible and that such a view deserves further philosophical and analytical attention. Moreover, although the view that I defend is broad enough to encompass the main features of the philosophical debates associated with disease language, I do not develop this point in the present paper.

The Pragmatic Significance of Illness, Sickness, and Disease

A pragmatic theory of disease highlights the way that disease language reflects specific kinds of practical concerns related to human suffering and disability. So said, it is important to characterize what gives rise to disease language from the base phenomena of human suffering and experiences of disability or incapacity. A defensible view is that there is a loose, but nonetheless important, relationship between the language of illness, sickness, and disease, a relationship in which explanatory control and interest is greater at the level of disease language than at the level of illness and that concern for the amelioration of human suffering moves from isolated, individual responses to the plight of ill or suffering individuals to the social organization of ways to respond effectively. The modern development of effective social responses to sickness features the use of disease language within a practice of medicine that is characterized by a commitment to scientific methods. This commitment to scientific explanation has been incorporated into recent philosophical theories of disease that readily accept scientific explanation as a standard for developing a philosophically defensible theory of disease. I discuss some of these approaches in the next section. For the present, I simply note that a pragmatic account of disease concepts can accept scientific interest as one—perhaps, even as the paramount—interest among others that shape disease language. However, it is not willing to accept that scientific standards for theoretical reason are philosophically preferable to standards internal to the social practice of medicine itself. What is too often overlooked and too easily forgotten is that a fully adequate philosophical understanding of disease needs to account for its *practical* functions as well as the theoretical status of disease concepts. To do so, one needs to attend to the dynamic relation that exists between the language and phenomena of illness, sickness, and disease and not just the theoretical, explanatory function that disease concepts serve when viewed from the perspective of a certain theory of science.

Illness is a subjective feeling or experience of an individual. It is a state in which the individual either feels bad or can not perform some normal action. Being unable to perform this action creates distress for the individual. Illnesses take many forms that basically involve distortions in one's phenomenological experience of self and world and of one's intentional action. These distortions can involve a wide range of algesic (pain-related) and teleologic (related to the failure of normal functioning) experiences that ultimately affect one's experience of the common social world. Thus, illness ultimately affects the individual's experienced ability to engage in human action understood broadly.[3] Thus, a key feature of being ill is one's own subjective self-experience of pain or incapacities of various sorts. Being sick, in contrast, is an experience that has an intersubjective trajectory and significance that involves socially influenced expectations and typifications of illness.

To be sick involves implicit claims about oneself as well as claims made by others about oneself. Being sick, at least in twentieth century America, involves four complementary features. First, a person who is sick can not get well simply by an act of will. To be sick means to suffer in the sense that one is passive with respect to the illness. This feature allows sickness to be distinguished from malingering or other intentional states. Second, the sick individual is relieved of some responsibilities. We allow, for example, sick leave and commonly accept sickness as an excusing condition for poor performance, because we recognize that sickness befalls individuals and diminishes or distorts their capacity for autonomous action, thought, and feeling. Third, it is expected that a sick person should want to recover as quickly as is possible, because sickness is an undesirable state. So, fourth, a sick individual should seek and cooperate with health professionals who are experts in healing. Sickness is thus a recognized social role in which an individual's subjective experience of illness is socially validated and accommodated within altered role expectations. The sick role clearly has moral implications. The sick person is expected to want to get well and is expected to cooperate with those who provide care. Disease is one way that health professionals validate an individual's claim to be in the sick role. An implication of this is that one is socially regarded as being *truly* sick only if one has a disease.

Diseases are thus practical concepts that serve broad social purposes of validation as employed by health professionals. Because the goal of cooperating with health professionals is to secure a restoration of health, concepts of disease also refer to an explanatory framework that enables the health professional to bring expert judgment to bear on the sick person's complaints, but the practical function precedes the theoretical or explanatory function of disease concepts. In other words, the evaluative element of disease is basically tied both to the phenomenology of illness and sickness as well as the practical purpose of ameliorating the suffering and incapacity associated therewith. The goal of scientifically explaining disease in terms of causal laws is thus not a free floating intention, but is ultimately grounded in a historically mediated response to the universal human experience of illness and sickness. Thus, to view the scientific project of explaining disease as paramount tends to eliminate the wider context within which this project is located and must ultimately be understood. To fail to set disease within this wider frame of reference not only loses sight of the wider cultural and social context of health care, but the specific practical intentionality of scientific clinical medicine.

As Tris Engelhardt has pointed out,

> evaluation enters into the enterprise of medical explana-
> tion because accounts of disease are immediately focused
> on controlling and eliminating circumstances judged to be
> a disvalue. The judgments are in no sense pragmatically
> neutral. Choosing to call a set of phenomena a disease
> involves a commitment to medical intervention, the
> assignment of the sick role, and the enlistment in action of
> health professionals ...
>
> Commitment to the concept of disease presupposes that
> there are phenomena physical and mental which can be
> correlated with events of pain and suffering, so that their
> patterns can be explained, their courses predicted, and their
> outcomes influenced favorably. Further, the pain and suf-
> fering can not be the immediate outcome of circumstances
> which are directly the subject of free choice. They must
> result from psychological or physiological laws; that is,
> they must be open to statement in the form of laws not
> moral rules. Medicine is the application of scientific, not
> moral generalizations.[5]

Disease language is thus best regarded not as a distinctively
moral response to human suffering, but as a particular kind of
response organized around scientific laws. It is important to
remember that magic, incantation, religious ritual, and alternative
healing practices are also responses to the phenomena of human
illness and suffering that do not rely on the language of disease,
although they involve roles analogous to that of the sick role and
they involve quasi-professional or, at least, expert healer roles.

What first distinguishes the specifically medical use of disease language is the implicit claim to the possession of a distinctive expertise in treating sickness based on scientific knowledge.
It is worth pointing out, however, that the belief in the scientific
basis of medical practice developed historically and was accepted
by society long before any reasonable scientific basis for medical
practice actually emerged. The social belief in the scientific basis

of medical practice precedes its actuality and so helps us to see that practical purposes are central to the use of disease language in medicine. The scientific nature of disease language is thus founded on deep cultural and social beliefs that themselves involve evaluative as well as theoretical commitments.

These beliefs are just that, *beliefs*. They are not validated knowledge. To explain fully the basis of these beliefs, we would need to analyze medicine's role in the emergence of scientific thought in the modern period and we would need to relate this development to ideals of health associated with the modern view of the good life, which centrally includes the idea of progress. Although such an exploration is beyond the scope of this paper, I do need to stress the point that interest in scientific explana- tion—or, more accurately, a certain model of scientific explana- tion—is in itself a pragmatic interest or set of interests that can be accommodated within the framework of a wider, pragmatic theory of disease. My critical discussion of the science-dominated mod- els of disease explanation in the next section is thus motivated by the need to show not only that they ultimately marginalize con- siderations of value, but that they mistake the interest in scientific theory for the ultimately wider set of pragmatic interests that affect the actual employment of disease language in clinical medicine.

Given the historical and socially accepted belief in medicine's commitment to science, it is understandable that philosophical treatments of disease concepts and theories have inclined toward the view that disease language is primarily explanatory and that a proper understanding of diseases would involve reductively locating them in terms of what is more basic from an explanatory point of view.[6] Nevertheless, even these treatments have had to come to terms with the stark realization that disease concepts carry with them a unique capacity to order human action that is as much practical as it is scientific. This practical or clinical significance is reflected in the common notion that disease lan- guage involves normative claims and commitments that themselves require philosophical analysis and justification. How these normative claims are understood and accommodated theoretically, however, sharply divides thinkers.

Normativism and Psychiatric Disease

In this section, I turn to three exemplars of the scientific theory-focused approach to disease language and discuss how well they accommodate normative or evaluative aspects of disease concepts. One well known view of the nature of the evaluative or normative features of disease language is represented in the criticism of the medical model in psychiatry and of the political use/misuse of psychiatric diagnosis that developed during the 1960s and 1970s. These critics argued that the language of mental illness and mental disease is a myth or label used to stigmatize nonconformist individuals.[7] These critics do recognize that disease language has a function in enjoining and justifying action with respect to sick individuals, but they deny that the relation between sickness and medical disease holds in a relatively similar way in the case of mental illness and psychiatric disease. In the more usual, medical cases, they argue, a person's illness and sickness are explainable in terms of value-neutral disease entities or processes that tie the experienced symptoms and complaints to a physical basis that comprises the disease. A medical (or true) disease is thus a descriptive concept that ties human experience to the laws of human biology or physiology. In all such cases of true disease, there is a physical basis for the disease that explains the sickness without relying on evaluative judgment. Of course, this overstates the relationship a bit, because the experience of sickness and illness on which the disease concept rests phenomenologically involve evaluative judgments for which one also needs an account, and this view fails to acknowledge that even in cases of so-called medical disease, the etiological element is often not fully understood. In the least nuanced views, the linkage between evaluative judgment and explanation is denied and value considerations are either ignored or explicitly expunged; other theorists, however, go to considerable lengths to acknowledge the contribution of evaluative judgment, but usually attempt to accommodate it in ways that maintain that disease language itself is really descriptively based and, hence, value-free.

Even though there are considerable differences in the degree to which the value aspect of disease, namely its basic sense of *dis*value, is accommodated within these theories, they hold what I have termed a *strong* normativist view of the function of values in concepts of disease.[8] The most extreme view is espoused by Thomas Szasz, who claims that disease always and necessarily involves a scientifically determined, and value-free, physical basis so that the only true diseases are medical diseases. So-called psychiatric diseases or mental illnesses are simply terms that hide social and value judgment within putative claims to scientific objectivity. Because the claims for a scientific basis of mental illness or psychiatric disease are not validated scientifically, their usage is treated as an example of political or social oppression of nonconformist or eccentric individuals. Thus, normativism in this strong sense is the view that psychiatric disease language is *purely* evaluative and involves no descriptive, i.e., empirical, basis at all. No wonder this view was regarded as a profound critical attack on psychiatric disease concepts and the social and political practice of psychiatric diagnosis.

Mental illness is regarded as a myth, because it imports social and moral values into the concept of disease that should be understood in scientific, objective terms as value neutral. What masquerades under the rubric of psychiatric disease are only "problems in living," not true disease states. Mental illness is a myth because true disease is physical in nature and amenable to scientific, objective description. It is thus value-free. Talk of "psychiatric" disease or mental illness is really a covert way of expressing moral or social judgments of praise or blame, of value or disvalue for what are essentially problems in living and not medical disease.

A philosophical corollary of the strong normativist critique of psychiatric disease is the so-called functionalist theory of disease, in which the concept of disease (as well as health) is defined descriptively in terms of those functions that are typically found within members of a species. Functionalism holds that disease is a descriptive and theoretical concept that is nonevaluative. Although not intended as a critical rejection of the social and political misuse of disease language (psychiatric or otherwise), this view limits the degree to which evaluative judgment can play a role in the language of disease. In fact, the leading proponent of the functionalist theory, Christopher Boorse, reverses the relationship between disease and illness on this point of values. In his view, "a disease is an *illness* only if it is serious enough to be incapacitating, and therefore is (i) undesirable for its bearer; (ii) a title for special treatments; and (iii) a valid excuse for normally criticizable behavior."[10] *Illness* thus introduces the evaluative element into the purely theoretical, scientific, and descriptive concept of disease. In this theory the value-neutral nature of disease is secured as a *theoretical* concept that is grounded in the concept of an atypical functional deficiency, namely, a defect that interferes with one or more functions typically performed within members of a species; and illness is treated as a *practical* concept that specifies diseases as particular disvalued states of affairs from a human, social point of view. Illness thus carries with it notions of disvalue, whereas disease is considered a value-neutral, descriptive concept.

Another recent example of a strong normativist point of view is Jerome Wakefield's application of the functionalist theory of disease to contemporary classifications of psychiatric disease.[11] This example is particularly important, because it incorporates a considerably nuanced understanding of the actual processes by which psychiatric diseases are constructed and classified and involves a conceptual analysis of many of the key elements involved in psychiatric disorders.[12] The most recent *Diagnostic and Statistical Manuals of Psychiatric Disorders* (DSM-III, DSM-

III-R, and DSM IV)[13] have cultivated considerable interest in the problems of psychiatric diagnosis and classification of psychiatric disorders and have had far-reaching effects on psychiatric practice and research.[14] The DSM-III/III-R definition of disorder revolves around two major points; first, that disorder has negative consequences for the person afflicted, and the second, that disorder involves a dysfunction in the person.[15]

Wakefield has developed an approach to the concept of disorder in the DSMs that carries out the strong normativist agenda. He rightly points out that the concept of disorder involves a practical or evaluative element, namely, that it is supposed to identify only those conditions that are undesirable and call for corrective social concern. This point, he argues, is expressed in the criterion of *harm,* which he identifies as a value-laden concept and which he thinks is the important source and kind of disvalue associated with the notion of a psychiatric disorder. Because Wakefield wants the concept of disorder to be grounded objectively (in order to distinguish disorders from other disvalued conditions), he argues that a secure ground can only be provided by what he calls the "scientific fact" that a dysfunction is a condition in which some internal mechanism is not working the way that it was naturally designed to work, as judged from an evolutionary biological perspective. In Wakefield's view, a major problem with the DSM-III/III-R definitions of disorder is their commitment to a value-neutrality that cannot be sustained within their own definitional constraints. This is because the DSM introduction of the idea of *negative* or *harmful* consequences uses value terms in the definition and thus embeds the definition and use of the concept in an evaluational realm. Wakefield regards this as the source of the sociopolitical debate about diagnosis, for example, the mid-sixties American Psychiatric Association homosexuality-as-disorder debate.[17] Such debate and discussion, Wakefield believes, is avoidable if one were able to achieve a theoretically-correct understanding of disorder.[18] Such an understanding is achieved by defining disorder as *harmful dysfunction:* dysfunction being the value-neutral and theoretical component and the harm component introducing the dimension of (dis)value.[19]

That the value dimension is important to nosological development and use is readily acknowledged by Wakefield,[20] but its role, in his view, needs to be restrained by a value-neutral, theoretical grounding of the notion of *dysfunction*.[21] Wakefield's solution is to define dysfunction along the functionalist lines sketched by Boorse, but with an emphasis on evolutionary biology. He defines dysfunction as "failure of a mechanism in the person to perform a natural function for which the mechanism was designed by natural selection.[22] To ensure the value-neutrality of the dysfunction/natural function concept, Wakefield further analyzes natural function as " ... an effect of the organ or mechanism that enters into an explanation of the existence, structure, or activity of the organ or mechanism."[23] Wakefield does admit that discovering what in fact is natural or dysfunctional may be extraordinarily difficult,[24] but he does not discuss this rather severe limitation in his account.[25] Instead, he appeals to a future scientific elaboration of mental function as the solution to this difficulty.[26]

For present purposes, it is important to note that these three accounts share a strong normativism with respect to disease concepts. They each believe that it is essential to differentiate descriptive or theoretical components in the concepts of disease from evaluative aspects and see the task of justifying disease concepts as primarily a task for scientific or theoretical reason. Advocates of these positions think that the concept of disease is a philosophically sound concept only if it (or some significant component) can be defined in value-neutral, scientific terms. Although this interest in justification is important, it tends to eliminate or push aside consideration of the place of justification within the wider, practical employment of disease language. The predilection for a theoretical account that is value-free thus greatly limits our understanding of the ways that psychiatric disease concepts are framed and used. It also, unfortunately, distorts the role of normative and evaluative aspects of disease language, because strong normativism involves a *separation* of values and scientific facts at least for purposes of philosophical justification.

Normativism involves attention to the normative or evaluative element in disease and does not itself entail a *separation* of values and scientific facts. Such a separation, however, is central to a strong-normativist critique of disease language. Values and facts are separated in strong normativist theories, because of a preference for a certain, logical empiricist, model of scientific explanation that is not at all required by normativism as such. For this reason, it is possible to outline an alternative normativism, one that is weak in the sense that it accepts that concepts of disease involve *both* descriptive and evaluative (or normative) elements and does not seek to exclude normative elements in order to attain a scientifically pure form of explanation.[27] A weak normativism is thus able to accommodate the pragmatic nature of disease language, the tension between theoretical or scientific demands for a descriptive basis for disease language, and the practical implications of the use of that language. Because the descriptive and evaluative aspects of disease language are inseparable, a weak normativism views descriptive and evaluative aspects of disease language as *components* of disease that illuminate broad areas in which philosophical analysis can prove useful, rather than restricting the range of philosophical discourse to the construction of a purely theoretical or scientific model of disease.

The perception that there is a serious conflict between an evaluative and a descriptive point of view is partly attributable to the dominance of scientific ways of thinking. As Bill Fulford has pointed out, from a science-based viewpoint, our classifications of disease appear to be predominately descriptive, because they are largely scientific in nature.[28] Viewed in these terms, values contaminate scientific explanation and so must be expunged. From a value-based perspective, however, all value terms carry descriptive as well as evaluative connotations so the philosophical task involves understanding the relationship between these terms rather than eliminating one or the other. Some value terms can have predominately descriptive connotations in contexts in which the descriptive criteria for the value judgments that they express are largely settled or agreed on.[29] Where convention or social agreement exists, value terms can be accepted as descriptions, but whenever disagreement occurs, the descriptive content tends to fall away and we are left with disputed evaluative content. From this point of view, then, it is easy to see why the evaluative connotations of psychiatric disease or mental illness are such an important source of debate, because the criteria are generally unsettled and so they appear largely as social/evaluative, rather than like the criteria of so-called genuine diseases as biological/scientific. As Fulford points out, physical pain is in most contexts and for most people an evil, whereas anxiety, a typical symptom of mental illness, is alternatively avoided by some people yet sought out by others; for example, when seeking the thrill of fear.[30] No wonder, then, that mental illness in which anxiety plays an important part will appear to be more value laden and, hence, more controversial than physical illnesses constituted by pain.

From a science-based point of view, disease appears factual because the concept is explanatory and articulated in descriptive terms, whereas illness appears to be evaluative, namely, disvalue appears more at home at this level. This point helps us to see how a science-based or theoretical orientation to the question of disease narrows our understanding by not developing the systematic relation between illness, sickness, and disease. A purely value-based view, however, can also constrain our understanding, because it inclines one to argue that both illness and disease are evaluative concepts that can be adequately explained by reference to a general process of social construction.

As a result, the pragmatic set of interests implicit in the practice of clinical medicine are simply treated as external political or social in nature, but not as specifically psychiatric or therapeutic interests internal to medicine as such.[31] What is missed is the distinction between two levels at which medical and psychiatric concepts can be considered: At the first level, illness and disease are treated as value terms *simpliciter* and at the second level, illness and disease are treated not just as value terms simpliciter, but as a particular kind of value, namely a *medical* as distinct from moral and esthetic values. Using this distinction, Fulford has pointed out that very little of the debate on the concept of disease has actually focused on the definition of medical value.[32] Examined at this second level, the actual experiences of illness, not as explained by disease concepts, but as phenomenologically given, are best analyzed in terms of a failure of action rather than as a failure of function. So regarded, pathology has to be understood not primarily by reference to disturbances of function, but primarily by reference to the patient's experience of incapacity or failure of one kind or another of ordinary intentional action. Such action failure models of disease have been developed in recent philosophy of medicine.[33] In cases of psychiatric disorders that are defined by failures of high-level capacities, such as, perceiving and believing, an action failure rather than a function failure account becomes necessary.[34] Actions, however, are in part actually defined by the values and beliefs of those performing them.[35] Pathology with respect to these high-level capacities are always failures of action and so can not be defined separately from values and beliefs of the subjects. Thus, psychiatric disorder turns out not to be a purely objective, descriptive notion after all, but (at least with respect to these high-level capacities) to be essentially embedded in the framework of the personal values and beliefs of the individuals concerned. The basic context of interpretation is, therefore, the context of human action, a context full of competing and contrary kinds of interests and purposes, rather than the context of biological function:

> Although less familiar than the analysis of disease as failure
> of function, the analysis of illness as failure of action has
> considerable face validity. First, as Toulmin (1980) pointed
> out, the very word *patient* implies a loss of agency. Second,
> it provides the required logical link between illness and

negative value. To intend something is to value it positively. Hence, a *failure* of intentional action is an inherently negatively evaluated experience. Third, and returning now to psychiatric classification, it is consistent with, and thus explains, the presence in ICD and DSM of a whole range of concepts which, if not exclusively the concern of philosophers, certainly sit as comfortably within the philosophy of action as with experimental science.[36]

A corollary of this view is thus a pragmatic understanding of disease that situates the employment of disease language within specific clinical, historical, and social contexts in which human action is expressed and experienced. The theory is pragmatic, then, in the sense both that it is concerned with practical matters and that its subject-matter focuses on distortions or frustrations in the range of typical human action that constitute illness.

The Pragmatic Goals of Care, Cure, Control, and Communication

A pragmatic understanding of the concept of disease offers two distinct advantages over other approaches. First, it allows and encourages a historically and socially situated analysis of the construction and use of disease concepts. For example, I have argued that a pragmatic approach to DSM-III-R offers a way of understanding the apparent contrary, if not conflicting, tendencies implicit in its overall conceptual framework as well as in the treatment of specific disorders, such as antisocial personality disorder.[37] Such an approach to specific disease concepts or nosologies encourages a rigorous, critical analysis of these concepts by requiring that the practical interest guiding the development, use, and justification of these concepts are made explicit before the task of philosophical justification is undertaken. Indeed, such an approach recasts the philosophical project into an engaged critical enterprise that is conversant with the clinical and practical employment of disease concepts rather than an abstract and idealized analysis of how these concepts *should* be used.[38] Thus, a pragmatic approach to the analysis of disease would press for a contextually sensitive treatment of disease concepts and theories.

Besides forcing a degree of concreteness into the philosophical understanding of disease, a pragmatic approach has the advantage of requiring a self-critical theory of disease. Such a theory would have to come to terms first with the systematic use of disease language in both clinical and nonclinical contexts. It would include an analysis of the relationship between illness, sickness, and disease like the one discussed above, relating disease concepts to a wider set of social and human interests. It would also attempt to identify and clarify the relation between various pragmatic goals that guide the construction of disease concepts. These are *general* goals. Identifying them helps to systematize our understanding of disease, but does not promulgate theoretical norms for the use of disease concepts. Rather, a pragmatic understanding of disease first attempts to identify and classify the norms and values that are implicit in the use of disease language within the practices in question and only secondarily to critically assess their employment. These rules operate more like rules in a language game or a practice than like formal rules of grammar or formal laws. Nevertheless, it is important to recognize that a pragmatic approach to the understanding of disease does afford a general way to organize the various functions and purposes that disease language serves and how this language expresses the historically determined practice. For convenience, we can speak of four general and distinctive pragmatic goals or purposes in the use of disease concepts: care, cure, control, and communication.

Concepts of disease serve the purpose of care by providing a basis for reassuring that the sick individual is ill not because of moral deficiency or through an act of will, but because of processes beyond direct control of the individual. Because the illness is understandable in terms of a concept of disease, the individual is assured and, so, is afforded a basis for placing his trust in the caregiver and for hoping for a restoration of well being. The goal of cure is similarly served by the explanatory nature of disease. In so far as disease processes are understood scientifically, it is possible to better direct therapeutic efforts to the pathophysiological processes associated with the disease. The concept of disease thus helps to differentiate the symptoms, signs, and underlying pathophysiological processes and events and provides direction for therapeutic corrective action. The ideal, of course, is to eradicate the disease whenever possible and to restore the sick individual to the full sense of well being that existed prior to the onset of the illness. Although cure may not always be possible, concepts of disease nonetheless direct care in ways that are supposed to be therapeutically successful. Here, "therapeutically successful" means that the distressing aspects of the disease process can either be eliminated or reduced even when the disease itself can not be eliminated. Thus, disease concepts afford a measure of *control* even when they are not able to effect cure.

The concept of control requires that a proper concept of disease will include pathophysiological understanding in addition to etiological components that afford health professionals the ability to isolate the disease within an individual and to apply measures to alleviate the symptoms of the disease. Controlling a disease can occur even when cure is not feasible. Chronic diseases, such as arthritis, can sometimes be controlled, namely, the pain ameliorated and the inflammatory and destructive processes slowed, even when the disease itself can not be eliminated. Because diseases exhibit patterns and these patterns afford a basis for prediction about the occurrence of diseases in individuals not presently affected, concepts of disease allow for health planning and public health measures by means of education, screening programs, or modifications of the environment.

Finally, disease language has the important function of promoting communication. Communication is foundational, because science is essentially a communicative process. Theories of disease that rely on the explanatory and scientific nature of disease language tacitly focus on this pragmatic purpose, but usually in isolation from other purposes and so tend to hide the rich complexity of communication that is involved in the scientific enterprise. Accordingly, a pragmatic theory of disease is able to incorporate the interest in scientific explanation within a wider context of interpretation. Theories of disease that highlight the scientific and theoretical aspects of disease language do so in a way that idealizes the scientific process. As a result, they fail to account for the actual historical and social processes that influence scientific investigation. In Hanson's terms, they focus on the logic of justification rather than logic of discovery.[39] Excessive attention to the demands and structure of scientific explanation, however, has created difficulty not only with respect to philosophical theories of psychiatric disease, but is also evident in the way that classifications of psychiatric disease have developed in recent years. In fact, the interpretive advantage of a pragmatic approach to the philosophical analysis of disease can be best illustrated by focusing on the problems that arise when psychiatric classifications of disease tacitly assume an idealized understanding of the scientific, descriptive nature of disease language.

Although the construction and classification of diseases serves the four goals of care, cure, control, and communication, these are not the only pragmatic goals and purposes that influence the process of constructing and classifying diseases. Other particular concerns related to the current state of scientific knowledge and the clinical problems currently challenging medicine can dramatically affect the way a disease is conceptualized. As Engelhardt has pointed out, the types of evaluative judgments that are involved in the selection of various clusters of phenomena as illnesses are diverse. They can be classified into three groups: the teleological, the algesic, and the aesthetic, each of which is concerned with a different aspect of human well-being.[40]

Further, diseases are not only multifactorial, but multidimensional, involving genetic, physiological, psychological, and sociological aspects. As a result, diseases have a basically relational character. For example, diseases like asthma, coronary artery disease, and arthritis are as much psychological and sociological as they are pathophysiological in the sense that the occurrence of illness is closely bound to experienced stress and the availability or lack of support for the person stressed. Thus, pragmatic interests affect not only what is counted as a disease, but more importantly how the disease concept is framed.

Conclusion

A pragmatic theory of disease is far more sensitive to the practice within which disease concepts arise and are used. A pragmatic theory first develops a concrete understanding of the practice and the implicit rules, beliefs, and values that shape a particular employment of disease concepts, and only secondarily engages in their critical analysis. Because the norms of a practice are internal to the practice, critical analysis of the use of disease concepts depends on an adequate understanding of the historical and social forces that shape the interests guiding the development, classification, and use of disease concepts. Interest in the scientific theory of disease is therefore a corollary of a wider set of interests with which a pragmatic theory of disease is concerned and not an exclusive preoccupation. Consequently, a pragmatic approach to disease language is attentive to the explanatory and evaluative elements of disease concepts, because they are essential features of the use of disease language in the practice of medicine. Recognition of the essentially dual nature of disease language is thus based on an understanding of the relationship between the phenomena of illness and sickness on the one hand and the use of the language of disease within the practice of medicine on the other.

Notes

[1]Temkin, O. (1961), The scientific approach to disease: specific entity
and individual sickness, in *Scientific Change,* Crombie, A. C.,
ed., Heinemann, London, pp. 629–647, argued that diseases exist
not as things in themselves or as ideal types, but as patterns of
explanation:

  The question: Does disease exist or are there only sick
  persons? is an abstract one and, in that form, does not allow
  a meaningful answer. Disease is not simply either the one
  or the other. Rather it must be thought of as the circumstance requires. The circumstances are represented by the
  patient, the physician, the public health man, the medical
  scientist, the pharmaceutical industry, society at large, and
  last but not least, the disease itself.'

  Such a contextualist account of disease has played an ever
important role in the history of concepts and theories of disease.
As Engelhardt, H. T. (1975), The concepts of health and disease,
in *Evaluation and Explanation in the Biomedical Sciences,*
Engelhardt, H. T. and Spicker, S. F., eds., D. Reidel Publishing,
Dordrecht, Holland, pp. 125–141, has pointed out, physiological
or functionalist theories of disease argue against the logical mistake
of confusing abstract concepts with things and treating them as
entities, but they do not deny that there are patterns of disease
processes; generally, such theorists adopt views in which diseases
are regarded as particular deviations from general regularities,
such as the laws of physiology.

[2]Canguilhem, G. (1978), *On the Normal and the Pathological,* D. Reidel
Publishing, Dordrecht, Holland, p. 11.

[3]Agich, G. J. (1995), Chronic illness and freedom, in *Chronic Illness:
From Experience to Policy,* Toombs, S. K., Barnard, D., and
Carson, R. A., eds., Indiana University Press, pp. 129–153.

[4]Parsons, P. (1975), The sick role and the role of the physician
reconsidered. *MMFQ/Health Soc.* **53,** 257–278.

[5]Engelhardt, H. T. (1975), Evaluation and Explanation in the biomedical
sciences, in *Evaluation and Explanation in the Biomedical
Sciences,* Engelhardt, H. T. and Spicker, S. F., eds., D. Reidel
Publishing, Dordrecht, Holland, p. 137.

[6]Whitbeck, C. (1977), Causation in medicine: the disease entity model. *Philos. Sci.* **44,** 619–637.

[7]Cooper, D. (1971), *Psychiatry and Anti-Psychiatry, Ballentine Books,* New York; Laing, R. D. (1967), *The Politics of Experience,* Ballentine Books, New York; Szasz, T. S. (1970), *The Manufacture of Madness,* Harper & Row, New York; (1970) *Ideology and Insanity,* Doubleday, New York; and (1961) *The Myth of Mental Illness,* Harper & Row, New York.

[8]Agich, G. J. (1994), Evaluative judgment and personality disorder, in *Philosophical Perspectives on Psychiatric Diagnostic Classification,* Sadler, J. Z., Schwartz, M. A., and Wiggens, O. P., eds., Johns Hopkins University Press, Baltimore, MD, pp. 233–245.

[9]Boorse, C. (1975), On the distinction between disease and illness, in *Philosophy and Public Affairs* **5,** 49–68 and (1982) What a theory of mental health should be, in *Psychiatry and Ethics,* Edwards, R. B., ed., Prometheus, Buffalo, NY, pp. 29–49.

[10]Boorse, C. On the distinction between disease and illness, p. 61.

[11]Wakefield, J. (1992), Disorder as harmful dysfunction: a conceptual critique of DSM-III-R's definition of mental disorder. *Psychological Review* **99,** 232–247 and (1992) The concept of mental disorder: on the boundary between biological facts and social values. *Am. Psychol.* **47:3,** 373–388.

[12]Sadler, J. Z. and Agich, G. J. (1995), Disease, functions, values, and psychiatric classification. *Philos. Psych. Psychol.* **2,** 219–231.

[13]American Psychiatric Association (1980), *Diagnostic and Statistical Manual of Mental Disorders,* 3rd ed., American Psychiatric Press, Washington, DC; (1987) *Diagnostic and Statistical Manual of Mental Disorders,* 3rd ed., Rev., American Psychiatric Press, (Washington, DC; (1994), *Diagnostic and Statistical Manual of Mental Disorders,* 4th ed., American Psychiatric Press, Washington, DC.

[14]Mezzich, J. E. and von Cranach, M., eds., (1988), *International Classification in Psychiatry: Unity and Diversity,* Cambridge University Press, New York; and Tischler, G. L., ed., (1987), *Diagnosis and Classification in Psychiatry: A Critical Appraisal of DSM-III,* Cambridge University Press, New York.

[15]It is worth pointing out that architects of the DSMs decided to substitute the concept of disorder for the concept of disease. This was a strategic substitution in their minds, because it allowed them to bypass the difficult debate about what constituted specifically *psychiatric* disease on the one hand and the inevitable confrontation of theoretical perspectives underlying competing views of psychiatric disease.

[16]Wakefield (1992), Disorder as harmful dysfunction: a conceptual critique of DSM-III-R's definition of mental disorder. *Psychol. Rev.* **99,** 237.

[17]Bayer, R. (1981), *Homosexuality and American Psychiatry,* Basic Books, New York; Politics, science, and the problem of psychiatric nomenclature: a case study of the american psychiatric association referendum on homosexuality, in *Scientific Controversies: Case Studies in The Resolution and Closure of Disputes in Science and Technology,* Engelhardt, H. T. and Caplan, A., eds., Cambridge University Press, Cambridge, pp. 381–400.

[18]I do not believe that sociopolitical debate or ethical analysis and controversy are necessarily undesirable aspects associated with the construction of psychiatric nosologies. However, I have argued that the DSM project would actually avoid a good deal of controversy if it were explicitly to embrace and analyze the contribution of evaluative judgment in its classificatory decisions. Agich, G. J. (1994), Evaluative judgment and personality disorder, in *Philosophical Perspectives on Psychiatric Diagnostic Classification,* Sadler, J. Z., Schwartz, M. A., and Wiggens, O. P., eds., Johns Hopkins University Press, Baltimore, MD, pp. 233–245. Doing so might help to achieve one of Wakefield's own implicit objectives, namely, that the discussion be rational and scientifically oriented toward questions of evidence rather than conducted as a rancorous political dispute. Even so, it would not eliminate the critical and rational analysis, discussion, and dispute about the value aspects and commitments of diagnostic categories.

[19]Believing that normativism entails a *separation* of values and scientific facts, however, overlooks another alternative, which I have termed a *weak normativism.* A weak normativism is the view that concepts of disease essentially include both evaluative and descriptive elements. It is a *normativism* in the sense that it recognizes that disease language enjoins action and reflects specific value

commitments. It is *weak* normativism, because it does not hold that the admission of evaluative elements into concepts of disease need eliminate the possibility that disease also includes a descriptive component. Weak normativism is a central commitment of a pragmatic approach to disease, because it forces us to keep in view the practical use of disease language, which always involves both evaluative and descriptive meanings. Because this form of normativism essentially accepts that concepts of disease involve both descriptive and evaluative (or normative) elements, it encompasses a far wider range of concerns than do competing accounts and, in fact, allows one to incorporate the philosophical interest in the specific *theoretical* aspect of disease as a preference for a certain kind of explanation. A good example of this analysis is Margolis, J. (1994), Taxonomic puzzles, in *Philosophical Perspectives on Psychiatric Diagnostic Classification,* Sadler, J. Z., Schwartz, M. A., and Wiggens, O. P., eds., Johns Hopkins University Press, Baltimore, MD, pp. 104–128.

[20]Wakefield, Disorder as harmful dysfunction, ibid. and The concept of mental disorder, ibid.

[21]This concern is the same one faced by Spitzer and Endicott, who used similar concepts to define disorder in terms of dysfunction (or abnormality, or maladaptiveness), namely, that the definition is circular unless one can secure a foundation for the concept of dysfunction that itself is independent of evaluative concepts. *See* Spitzer, R. L. and Endicott, J. (1978), Medical and mental disorder: proposed definition and criteria, in *Critical Issues in Psychiatric Diagnosis,* Spitzer, R. L. and Klein, D. F., eds., Raven, New York, pp. 15–40.

[22]Wakefield, (1992), Disorder as harmful dysfunction.

[23]Wakefield, (1992), The concept of mental disorder, p.382.

[24]Wakefield, (1992), Disorder as harmful dysfunction, p. 236.

[25]Sadler and Agich, (1995), p. 222.

[26]Sadler and Agich, (1995), p. 227.

[27]Agich, (1994), Evaluative judgment and personality disorder.

[28]Fulford, K. W. M. Closet logics: hidden conceptual elements in the DSM and ICD classifications of mental disorders, in *Philosophical Perspectives on Psychiatric Diagnostic Classification,* Sadler, Wiggins, and Schwartz, eds., pp. 211–232.

[29]Fulford, ibid., p. 217.

[30]Fulford, ibid., p. 219.

[31]*See* Agich, G. J. (1983), Scope of the therapeutic relationship, in *The Clinical Encounter*, Earl E. Shelp, ed., D. Reidel Publishing, Dordrecht, Holland, pp. 233–250.

[32]Fulford, ibid., p. 221.

[33]Nordenfelt, L. (1987), *On the Nature of Health: An Action-Theoretic Approach,* D. Reidel Publishing, Dordrecht, Holland; and Fulford, K. W. M. (1989), *Moral Theory and Medical Practice,* Cambridge University Press, Cambridge, Chapters 7–10.

[34]Fulford, Closet logics. Chapter 10.

[35]White, A. R., ed. (1968), Introduction, in *The Philosophy of Action,* Oxford University Press, Oxford.

[36]Fulford, ibid., p. 222.

[37]Agich, G. J. (1994), Evaluative judgment and personality disorder, ibid.

[38]A good example of the richness of this approach is to be found in the chapters collected in the Sadler, Wiggins, and Schwartz book, *Philosophical Perspectives on Psychiatric Diagnostic Classification.*

[39]Hanson, N. R. (1958), *Patterns of Discovery,* Cambridge University Press, Cambridge, UK.

[40]Engelhardt, H. T. (1976), Human well-being and medicine: some basic value-judgments in the biomedical sciences, in *Science, Ethics, and Medicine,* Engelhardt, H. T. and Callahan, D., eds., Hastings Center, Hastings-on-Hudson, New York, p. 132.

[41]Engelhardt, H. T. (1974), The concepts of health and disease, in *Evaluation and Explanation in Biomedical Sciences,* Engelhardt, H. T. and Spicker, S. F., eds., D. Reidel Publishing, Dordrecht, Holland, p. 33; and Holmes, T. H. and Rahe, R. H. (1967), Social readjustment rating scale. *J. Psychosom. Res.* **11,** 213–218.

Introduction

This article discusses a number of disease definitions, particularly those of Christopher Boorse, and argues that "disease" permits a rough characterization but ultimately resists definition. Although it seems correct to highlight the pathophysiological dimension of disease as "essential" to its comprehension, as Boorse does, Banja argues that his injecting concepts like normalcy or functionality into that definition invite subjective or cultural notions to infiltrate the meaning of disease. Nevertheless, these cultural artifacts seem inescapable since in order to understand the "pathos" of pathophysiology, some value-laden notion of deviance, atypicality, abnormality, or undesirability is necessary. Banja believes that the key to characterizing (but not defining) disease consists in the way that a society's health professionals relate their understanding of pathophysiologic processes, subjective symptomatology, and objective clinical signs to cultural attitudes about the threat to human welfare that is posed by a health related adversity and what measures are socially and professionally recognized in remediating the problem. Banja discusses Richard Rorty's antiessentialist approach, which suggests that our understanding of terms like "disease" is mediated by the way our beliefs about their referents cohere instead of our having discovered that singular essence that accounts for their "being." Furthermore, if Banja's account is correct, a definition of disease will never be terribly important for practical purposes since clinical praxis will be more oriented to the relief

247

of suffering regardless of whether that suffering is labeled a disease, malady, impairment, syndrome, ailment, disability, or whatever. Diseases are bound together not by an essence articulated through definition, but rather by notions about disease that make up what Rorty calls a "web" of beliefs as well as structural commonalities that Wittgenstein described in his famous account of family resemblances.

Defining Disease

Praxis Makes Perfect

John D. Banja

In their classic *Textbook of Medicine,* Beeson and McDermott (1975)[1] do not index "disease." Instead, one finds: "Disease. See names of specific diseases." Whether the authors are ontological pluralists or just theoretically indifferent, the absence of a specific entry on disease reminds one of Hesslow's recent observation that only philosophers, social scientists, and public health officials evince a keen interest in defining disease.[2] Clinicians and medical scientists have never appeared particularly upset over the absence of a refined, universally accepted definition of a disease; indeed, they seem to get along splendidly without one.

Why should a definition that expresses the essence of disease be possible at all? On the one hand, diseases get identified or named in at least four, rather unrelated ways:

1. After the discoverer (Alzheimer's, Parkinson's, Kreutzfeld-Jacob);
2. By way of the anatomic site that is affected (hepatitis, bowel/bladder/bone cancer, coronary artery disease, retinitis pigmentosa, polycystic kidney disease);

3. Vis a vis the cause of the disease (silicosis, chronic beryllium disease, and so on);
4. By those afflicted with the disease (e.g., Lou Gehrig's Disease, Legionnaire's disease, miner's asthma, mushroom picker's disease, bridegroom's disease).

Some diseases, as I shall argue later, ought not be called diseases at all. On the other hand, "disease" can refer to the complaints that sick people present to their physicians, explanations regarding why people get sick, excuses from responsible behavior ("I'm sick, so I will not go to work"), entitlements to certain social benefits, such as Medicare reimbursement for dialysis, and justifications for otherwise unallowable intrusions by the government on individual liberty, such as forced immunizations in time of plague.[3]

Mersky has stated flatly that "No agreed definition of disease exists."[3] Even so, it is important to refine our understanding of disease for the obvious reason that doing so might tell us something important about ourselves and how we conceptualize our afflictions. (It can also suggest some criteria for decisions about allocating health care, although that topic will not be discussed in this paper.) I will comment on certain characterizations and definitions of disease, especially those of Christopher Boorse, and propose that a fairly reliable understanding or model of disease emerges. This model stops considerably short of a definition but constitutes, I think, about the best we can do. I will simply say, very much as Engelhardt noted in 1978,[4] that disease is an organizing concept or conceptual model that embraces the way a pathophysiologic event links up with and explains observable signs or symptoms characteristic of "illness" that enables a rational, coherent, and reliable conceptual basis for a medical intervention in a given society.

Since that position is hardly novel, I will argue the more generic claim that "disease" is not definable at all, but that it exhibits some traits that are fairly prominent in what Richard Rorty might call our "web of beliefs" about disease.[5] I will employ a number of Rorty's recent criticisms of essentialist approaches and conclude that, at best, "disease" is an umbrella term that embraces a number of clinical entities or conditions that are related according to the Wittgensteinian notion of family resemblance.

Defining Disease

The obvious test of any disease definition is that it apply to and explain why its instantiations *really are* diseases. If the application/explanation demonstration proves unsatisfactory, we have two options: either abandon the definition or rename those phenomena formerly called diseases and, presumably, identify new ones. The latter, of course, leaves us with the problem of justifying the new classification system as nonarbitrary.

Given the fact that the plethora of events we call diseases are remarkably wide ranging and that we only have rough conceptual boundaries to differentiate disease from terms like syndrome, impairment, handicap, illness, disability, and the like, it may seem odd that health care providers—who, after all, are confronted with the challenge of treating medical afflictions and maladies every day—do not agitate for more conceptual precision in this terminological hodge-podge.

Like Christopher Boorse, I am attracted to those "realist" definitions of disease that look to biological functioning for their intelligibility. Boorse claims that diseases are what

> interfere with one or more functions typically performed within members of the species ... [D]eficiencies in the functional efficiency of the body are diseases when they are unnatural, and they may be unnatural either by being atypical or by being attributable mainly to the action of a hostile environment.[6]

In another article, Boorse (1977) notes that disease is:

a type of internal state which is either an impairment of
normal functional ability, i.e., a reduction of one or more
functional abilities below typical efficiency, or a limitation
on functional ability caused by environmental agents.

Boorse believes that this naturalistic account "frees the idea
of theoretical health of all normative content" because it refers us
to an objective realm of human physiology wherein biological
functions occur independently from human interests, values and
goals. Indeed, this objective realm of "deficiencies in functional
efficiency" allows him to distinguish disease from illness or at
least to categorize illness as a subclass of disease. For him, disease is a state of the organism that compromises physiological
functioning. Illness is disease that has been contextualized, i.e.,
it has been deemed undesirable and connotes a possible entitle-
ment to special consideration (e.g., medical treatment, relief from
social responsibilities, and so on).

Because Boorse's accounts are conceptually rich, they have
drawn much comment. I believe he is absolutely right to call
attention to the ubiquitous role that cellular or organic pathology
plays in our understanding of disease. On the other hand, two sets
of criticisms can be suggested. The first deals with conceptual
problems inherent in exorcising the specter of conventionality
from terms like atypicality, unnatural, and normal. The other set
of criticisms concerns the way Boorse's disease definition does
too much, such as in encompassing conditions like disabilities or
injuries. Such application seems wrong, because we do not ordinarily say that the person who is unable to speak because of a
stroke she experienced 10 years ago has a disease, nor do we say
that the individual who sprained his ankle playing soccer is diseased. Although both of these cases seem to fit Boorse's definition of disease, I shall argue that it is not simply a matter of
linguistic preference regarding whether or not disabilities or
injuries are disease; indeed, in a later part of this article I shall
argue they are not. For now, though, we might turn to problems
with Boorse's reliance on terms like normal and atypical.

First of all, as Temkin (1963)[8] observed, if we define disease only in terms of an underlying physiology—and understand that pathology is some kind of deviation from what is "normal"—then there would be as many different diseases as there are pathogenic organisms or abnormal molecular chemical structures. On that account everyone would be diseased in one way or another since all of us exhibit microscopic or macroscopic cellular or organic irregularities. However, Boorse notes that these irregularities must express themselves as deviations from customary functional efficiency, i.e., as deviations from both *what* a species normally does and the *usual efficiency* characteristic of that behavior or activity (i.e., the measurable, functional parameters that quantitatively represent the upper and especially lower limits of "normal" functional activity).

In criticizing Boorse's notion of what counts as normal vs abnormal functioning, however, Brown (1985)[9] calls attention to Ruse's concern that such a search for normalcy threatens circularity. That is, to find out what is normal for X, I have to determine what Xs do, but how do I identify the population of Xs? Perhaps, as Boorse says, analysis of a sufficiently large sample of a population of Xs over time will yield a "species design" as a standard reference, but how does one calculate the scope or membership of that population? If we are speaking of humans, do we include newborns as well as nonagenarians, the dying and the chronically ill as well as the robust and healthy, and the disabled and functionally impaired as well as the able-bodied? Note that persons with disability—who would appear to be paradigmatic cases of physiologic dysfunction—are no small group since estimates of their number in America exceed 40 million (about one out of every six or seven citizens). Might our species selection from which we propose to derive our notion of normalcy be biased because we already have a notion of normalcy in mind that will be confirmed by the sample we have already preselected? Put another way, in order to distinguish a population of normals, I ought to have a concept of nonnormals, i.e., persons with disease, but if I am wanting to identify that population of normals in order to ascertain a definition of disease, I am caught in a circle, unable to justify that my group of normals is "really" normal because I am without a concept of nonnormalcy.

These criticisms lead to a deeper reservation I have about Boorse's position, which is that the only way to put real meaning into a notion of disease is to admit what he resists, namely, that disease-related notions of atypicality or unnaturalness rely on social or conventional understandings of "pathology, i.e., that the upper and lower limits of "diseased" functioning are socially determined. The point is that we do not become interested in atypical function as a "disease" until it interferes with our performing certain behaviors, or our subjective sense of well being, or as Whitbeck (1977)[10] observed, it becomes something we want to prevent. Our attempts to define disease—by way of identifying that transcendental something that constitutes its "essence"—get into trouble because, as others have noted, disease is an inherently and inevitably valuative term that has normative as well as descriptive connotations. It is contextually driven in that its understanding is inextricably tied to a culture's comprehension of "illness" and what health care measures, if any, should be undertaken when an illness appears. (This could very well explain a reluctance to call symptomless gallstones or lipomas diseases, whereas sterility in a densely overpopulated country would not be seen as a disease at all.) The line of demarcation between atypical, nondiseased function and diseased, pathophysiologic function is drawn as much by cultural attitudes toward pain and sickness as it is by the scientific comprehension of cell membrane permeability, electrolyte imbalances, and the presence of toxic microorganisms. These claims will be fleshed out in what follows.

An Antiessentialist Perspective

In his essay "Inquiry as Recontextualization," Richard Rorty (1991)[5] proposes that a society's concepts already come contextualized and that essentialist approaches—i.e., those that seek definitions—go wrong as they abandon that context and aspire to a "God's eye" view of reality (i.e., the way reality exists apart from human sentience, interests, beliefs, values, aspirations, and so on). Thus, the Western penchant to understand disease by way of materialistic reductionism is as much a cultural artifact as the Central American propensity to understand disease as resulting from hexes. In either case, though, the Rortian view I want to advance would say that our understanding of disease derives from a conceptual scheme whose persuasiveness does not depend on how well that scheme represents reality (which is certainly congruent with views like Boorse's), but rather on how well the beliefs contained in that scheme cohere with one another.

The cognitive structure that explains and integrates our beliefs about diseases is not mediated by an intuition of the essence of disease (as articulated by a definition) but derives from a learned recognition of patterns that link physiologic dysfunction, objective signs, subjective symptoms, and cultural values that determine interventions. For example, consider the recent discoveries that will doubtlessly confirm the designation of chronic obesity as a disease. Cellular biologists have recently located molecular "antenna" that receive messages from a protein in the blood, leptin, and pass a "satiation" message to the brain's fat burning and appetite control centers.[11] Although people who are obese manufacture leptin, data suggests their brains do not process the satiation message and therefore they continue to eat long after they should have stopped.

This seems to be a good example of a Boorsean disease: an impaired physiologic function that is unnatural. As a science-based account that looks to biologic functioning as it exists independently from human conventions, this account likewise seems to provide a powerful rejoinder to claims that obesity derives from gluttony. Understood pathophysiologically, obesity seems more on par with diabetes or hepatitis.

But what precipitated this research is precisely the cultural belief that obesity is undesirable—a valuatively driven notion that is hardly universally shared. For example, Galanti (1991)[12] described a situation wherein an obese woman met her Central American boyfriend's parents, who persistently and approvingly called attention to her weight. Her boyfriend later explained that his parents found her size aesthetically pleasing and socially valuable since, in their eyes, her weight would enable her to have many children.

Notice that even if obesity can be demonstrated to dispose one to a greater probability of chronic disease or a shortened life span, the undesirability of these situations may be offset in a particular culture by certain, deeply valued aesthetic dispositions. The scientific repudiation of obesity inheres in a valuative judgment that deems it an excessively "risky" condition. Yet, what counts as an acceptable vs unacceptable risk is clearly a sociocultural decision.

All this, however, only underlines the normativity of disease concepts:

> the inability to discriminate colors, to taste phenythiocarbimate, or to roll the sides of one's tongue inward may or may not be counted as diseases, depending upon the contribution that these functions make within a particular context or environment ... Definitions of illness and disease will (insofar as essential human functions cannot be identified and disease models are arbitrary) be dependent upon social conventions. If this is the case, it will readily follow that definitions of illnesses and diseases have deep social roots as well as broad social consequences.[4]

Essences, Prototypes, and Frames

Essentialists, of course, do not like this kind of talk since they want their beliefs to correspond to reality and portray it as it ultimately is. For them, if beliefs about reality cohere, it is only because reality itself hangs together and one's metaphysics reveal that underlying transcendental structure and system of relations that accounts for such coherence. But pragmatists like Rorty delight in pointing out that correspondence theories encounter immense problems in securing a nonarbitrary language by which to mirror that reality as well as related evidentiary problems rooted in the impossibility of leaving our bodies to compare how our beliefs measure up to reality.

It is beyond the scope of this paper to survey these arguments. Suffice it to say, though, that if diseases do share a common something (or, as Rorty might say, a "super" something) that explains their "diseaseness" and therefore differentiates them from nondiseases, I cannot imagine what it could be. The disease model of pathophysiologic dysfunction + signs/symptoms + medicocultural values encouraging or discouraging treatment interventions that I am advocating is only that—a model that clarifies what we ordinarily understand as disease. It does not, however, distinguish disease from injury or impairment, nor does it adjudicate hard cases, such as whether or not small stature, addictive behaviors, or insomnia qualify as diseases. Certainly, diseases do not share a common etiology since some are caused by microorganisms, others by toxins, others by a hostile but natural environment (e.g., gangrene from frostbite), and others from genetically based pathology. Some diseases resolve spontaneously; others, like amyotrophic lateral sclerosis, are thus far incurable. Some have a rapid onset, like the Ebola virus; others, like prostatic hypertrophy, can take years to evolve. Some diseases appear, run their course, resolve, and never reappear. Other diseases, like polio, appear, run their course, become dormant, and then often reappear many years later with considerable gravity. Some phenomena, like alcoholism, were once thought not to be diseases but now are; others, like drapetomania (i.e., the tendency of slaves to flee from bondage), were once termed diseases, but are now only embarrassments of medical history. Some diseases are one

with their manifestations, such as the cellular necrosis characteristic of gangrene. Other disease processes are distinguishable
from the conditions they cause, like diabetes causing renal failure
or blindness. Finally, some diseases are misnamed. Parkinson's
is more a syndrome than a disease since it manifests itself in
numerous ways (e.g., upper extremity or head tremors, slurred or
racing speech, and the characteristic "Parkinsonian gait"). Perhaps we call Parkinson's a disease because, like diabetes, its
etiology is so neatly circumscribed despite the symptom complex
that argues for its being classified as a syndrome.

Whereas Rorty repudiates essentialist approaches on philosophical grounds, Mark Johnson (1993)[13] assumes an antiessentialist position on neuropsychological grounds. He argues that
humans understand categories or terms not by discerning a platonic or husserlian essence, but through our familiarity with a
prototype:

> Psychologists, linguists, and anthropologists have discov-
> ered that most categories used by people are not actually
> definable by a list of features. Instead, people tend to define
> categories (e.g., bird) by identifying certain prototypical
> members of the category (e.g., robin) and they recognize
> other nonprototypical members (e.g., chicken, ostrich, pen-
> guin) that differ in various ways from the prototypical ones.
> There is seldom any set of necessary and sufficient features
> possessed by all member of the category. In this way our
> ordinary concepts are not uniformly or homogeneously
> structured.

Johnson also calls attention to the idea of "frame semantics," in which terms and concepts get their meaning from larger
cognitive structures or frames. Although there may be limits to
how encompassing a frame can be and the degree to which dis-
parate terms might be embraced by that frame, these limits cannot
be fixed in advance since they grow out of our experience. But
that experience is primarily determined by the collective experi-

ence of the language users in a community. To the extent that the meaning of an experience is governed by the "frame" we bring to the situation, the more frames that are available, the more varied those experiential meanings become.

The relevance of frames for the notion for "disease" is, I believe, that the model I am encouraging is not the only one possible. The primary impetus for our employing a physiologic element in characterizing disease is that pathophysiology has evolved, from roughly the beginning of the 19th century, as the dominant frame by which the West understands disease. Someday, we might employ a spiritual model—e.g., certain people would like to argue that "sin" is the ultimate cause of disease—or some other construct. Many individuals, for example, want to urge the acceptance of "somaticized diseases, for which no pathologic physiology is identifiable.[14] These "diseases" are controversial precisely because no convincing etiology is available to explain the complaints whose legitimacy and authenticity are not in doubt. Moreover, the recent interest in alternative methods of healing and the always interesting issue of the efficacy of placebos might very well result in other models of disease being placed alongside what I take to the prevailing model I am describing here.

What Would a Refined Definition Accomplish?

Hesslow (1993)[2] has recently wondered what a refined definition of disease would accomplish. What is the value of knowing how things really are "precontextually"?

It does not seem that a refined definition of disease would improve the quality of disease treatment because health providers are already disposed through the collective wisdom and knowledge of their profession toward what counts as "legitimate" signs and symptoms meriting treatment, whether or not these manifestations count as a disease. Imagine a physician saying, "Yes, you are clearly ill for obvious biophysiological reasons. However, what you have does not qualify as a disease by the accepted definition; therefore, I will not treat it."

On the other hand, a refined definition will not assist payers of health care to make better decisions about whether or not to reimburse the costs of disease care. Such decisions are currently made according to actuarial information, marketplace competition, and consumer demands. Some insurers will pay for bone marrow transplants for treating cancer, whereas others specifically exclude this coverage in the health insurance plan.[15] Clearly, the context of economics is what drives these decisions irrespective of what a refined definition of disease would include.

Does all of this mean, then, that our understanding of disease proceeds without a direction and that what counts as a disease is more a matter of fashion than science, taste rather than reason? If convention and social values are all that we have in understanding and managing disease, then why is it that the presence of fecal blood, swollen glands, or a blood pressure of 200/100 mm will immediately and universally engage the concern of the Western medical community? To say that disease requires "as its sufficient and necessary condition the experience of therapeutic concern by a person for himself and/or the arousal of therapeutic concern for him in his social environment"[16] implies what no one would accept: that the allocation of medicine is helpless in the face of complaints from hypochondriacal patients and the ministrations of medical scallywags who have never met a complaint they did not like.

What is it, then, that provides structure to the pragmatist's web of beliefs about disease, that explains why certain beliefs about disease cohere so well (i.e., occupy a central place in the web), and why we do not replace them at whim? In sum, what explains the fact that although a definition of disease remains elusive, health providers neither lament nor despair over its absence?

For Western health care professionals, the answer lies precisely in Boorse's notion of biophysiological dysfunction. We understand the objective signs and subjective complaints of some-one who is sick according to the degree to which some underlying cause can be determined: Objective physiology informs subjective symptomatology. The relationship is causal and even a self-avowed antiessentialist like Rorty (1991)[5] is ready to concede the primary role of causality in the formation of beliefs:

> … we shall not take ourselves to have found such a coherent pattern (of beliefs) unless we can see these (human) organ-isms as talking mostly about things to which they stand in real cause-and-effect relations.

Rorty will not succumb to the seduction of realism here, however, since he believes nothing can transport us from our web of beliefs into a realm of essences. Because that web of beliefs is all we have, Rorty would say that those beliefs attaching to biological reductionism are so fundamental to our understanding of disease that jettisoning them would entail a Kuhnian revolution in the comprehension of disease.

Disease, then, admits a ubiquitous and objective referent—one occupying a central place in Rorty's web—in terms of a pathologic physiological event. This is what urges realists like Boorse into arguing that a nonarbitrary definition of disease is possible. What I cannot grant Boorse's definition does is pro-vide us with an explanation of our ordinary use of the word disease that explains why certain types of atypical physiologi-cal function are not regarded as diseases, such as minor ailments like a sprained ankle or serious impairments like aphasia or hemiplegia following stroke. I believe his definition does not do this because it fails to respect the degree to which conventional medicine operating somewhat in tandem with social values reserves a special place for identifying a medical condition as a "disease." That special place cannot be captured by a defini-tion, but rather by appreciating how an assemblage of empirical and conventional factors link up and provide an operational meaning of disease. It is not enough to simply refer to atypicality or unnaturalness without knowing how a society interprets the nature and degree of the affected individual's condition and the degree to which the society is able to and interested in deploying certain therapeutic interventions.

This explains why, paradoxically, physicians do not need a definition of disease since their ministrations neither begin, end, nor depend on one. Rather, they employ observation and an inductive/synthetic approach that considers whether the complainant's signs and symptoms display a meaningful, recognizable pattern that can be associated with an "accepted" etiology and course of treatment. Researchers do not need a definition of disease because they can only understand the biological properties connected with disease in terms of other beliefs their own community of scientists has evolved. The beliefs of both physician and researcher cohere through the confluence of values, knowledge base, and problem-solving strategies of their respective communities. For them, coherence, not definition, is key.

The fallout of these remarks is that our world, as legal scholars have long recognized, is not the conceptually neat and tidy place that philosophers attached to the classical tradition seeking definitions for conceptual security would like it to be. Indeed, attempts to arrive at definitions in law tend to be as difficult but considerably more disconcerting than in philosophy, since the former must be operational whereas the latter remains largely confined to the esoterica of metaphysics and epistemology. Law professor Leo Katz (1987)[17] has noted that the law responds to the problem of operational definitions by trying to imagine every situation in which a term might occur, and then stipulate which conditions are covered by the term and which not. Thus,

> the Securities Act of 1933 opens with an interminable, thirteen part definitional section seeking to supply exhaustive definitions for all of the act's basic terms, from "security" ... to "person" ... The Uniform Commercial Code opens with a similar forty-six-part section defining everything from "aggrieved party" ... to "fault."

Even otherwise exemplary definitions have their problems. Although no one would gainsay the proposition that "All bachelors are unmarried males," are persons who have lived together for a year "married" and are men who have undergone feminization hormone treatments really women?

Disease, Disability, and Injury

I mentioned earlier that Boorse's definition of disease does not adequately distinguish disease from disability or injury, yet the latter conditions clearly owe their existence to some pathophysiological event that reduces the functional efficiency of an organ system. I believe its failure to do this is significant since we do not commonly say that the person with aphasia or an amputated limb has a "disease," but a disease definition should either explain or persuasively repudiate ordinary uses of language.

To examine the relationship among disease, disability, and injury, we might look to the behavior of the medical community in managing these conditions. If the clinical community seeks to diminish or reverse a pathologic process and aim at cure, i.e., realize a return to the individual's premorbid or preillness state, then we are ordinarily disposed—perhaps in a Johnsonian "prototypical" way—to think in terms of a disease. If the therapeutic emphasis is on assisting the individual to adapt to and cope with his or her malady or functional impairment, it is likely that we are in the presence of what would ordinarily be referred to as a disability.

A nontraumatically caused disability is always precipitated by a disease process, which may or may not continue to coexist with the impairment or disability it caused. For example, stroke is caused by either a ruptured or a blocked vessel that shuts off the blood supply to some part of the brain.[18] The affected neurons die off within minutes and are not replaced. People who recover fully from their strokes do not significantly manifest signs or symptoms, hence we neither say they are diseased nor disabled. Of those who do not fully recover, why don't we say at the time their disability is fairly established and permanent—usually about six months after the stroke—that they have a disease?

A prominent reason is that, as Whitbeck (1978)[10] observed, we prototypically think of disease as a process, not a state. Consequently, all strokes are preceded by a disease process causing the thinning or thickening of cerebrovascular walls or causing the formation of an embolus. Once a serious stroke occurs and the disease process leading up to it is presumably completed, permanent neurological damage has been done and is now manifested as some sort of behavioral, cognitive, or motor impairment. We do not call the impairment a disease if the causative pathophysiologic process has run its course and the impairment seems "stabilized." If it is suspected, though, that a disease process may still exist by way of a continuing atherosclerotic process that threatens another stroke, prophylactic treatment—most often at least consisting of antihypertensive medications—commences. We might say, therefore, that disability is what happens when an organ system that has been impaired by a disease process interfaces the world. The functional deficiency that characterizes that disability is a manifestation of a disease process, but separable from it since "deficiency" or "disability" must look to the understanding and interpretation of a society's members in order to gage its meaning and severity.

If we turn to distinguishing disease from injury, we become intuitively and practically aware that the two are different: A spinal cord injury is vastly different from a meningioma of the spinal cord, even though they might cause similar paralyses. Every physician appreciates the difference between disease and injury but disease theoreticians do not. Why? Because the difference between injury and disease is an empirical one, found in the microscopic cellular or biochemical changes that are characteristic of disease vs the usually macroscopic organic changes that occur from the traumatically induced force loading on organ systems. Physicians recognize the difference because they take a history, perform a physical, and order laboratory data. As they synthesize the data from these *multiple* sources, they differentiate diseases from injuries according to characteristic patterns that they learned in medical school, e.g., the presence of readily identifiable microorganisms characteristic of disease vs the conjunction of pain with edematous, inflamed, and contused tissue characteristic of injury. At base, diseases simply look different from injuries, but a disease definition cannot capture that difference since we cannot be logically certain that an injury will appear whose manifestations will mimic those of injury. Indeed, injuries in the workplace, such as reflex sympathetic dystrophy blur distinctions between disease and injury, especially if the pathophysiologic event evolved over a long period of time.[19] By and large, however, we appear to get along relatively well by just having learned to recognize the appearance of certain characteristic patterns, some associated with disease and others with injury. As Wittgenstein (1958)[20] encourages us in a famous passage, "Don't think, but look," and although no medical professor would encourage his or her students to abandon thinking, I have yet to hear one of them recommend that the key to an improved of understanding of disease is through an improved definition.

Conclusion

As Beeson and McDermott perhaps intuitively realized, our understanding of disease grows out of our experience with diseases, an experience that is ultimately marked by a collective struggle to relieve suffering. In any era, a few diseases will occupy center stage and suggest the disease prototype. The great physician Sir William Osler was said to have remarked, "Know syphilis in all its manifestations and relations and all other things clinical will be added unto you.[21] Tuberculosis and gastroenteritis were major killers in the early part of the 20th century, considerably outdistancing cancer. Today, our collective consciousness is taken up with AIDS, Alzheimer's, and coronary artery disease.

Diseases do not share an essence, but rather form a family, loosely associated by certain pathophysiologic events expressing themselves as signs and symptoms that a particular society deems undesirable. What is crucial, though, is that the *linkages* between pathophysiology, objective signs, subjective symptoms, and their deemed undesirability are culturally, not logically, determined. To know disease, one must know the ways his or her culture regards diseases—and that knowledge is derived experientially, not definitionally. Wittgenstein's famous passage on the essence of games can be applied equally well to diseases:

> What is common to them all?—Don't say: "There must be something common ... "—but look and see whether there is anything common to all.—For if you look at them you will not see something common to all, but similarities, relationships, and a whole series of them at that. To repeat, do not think, but look![20]

In matters of defining disease, I believe Wittgenstein's suggestion is the best we can do.

References

[1]Beeson, P. B. and McDermott, W., eds. (1975) *Textbook of Medicine,* W. B. Saunders, Philadelphia.

[2]Hesslow, G. (1993) Do we need a concept of disease? *Theor. Med.* **14,** 1–14.

[3]Mersky, H. (1986) Variable meanings for the definition of disease. *J. Med. Philos.* **11,** 215–232.

[4]Engelhardt, H. T. (1978) Health and disease: philosophical perspectives, in *Encyclopedia of Bioethics,* Reich, W., ed., Free Press, New York.

[5]Rorty, R. (1991) Inquiry as recontextualization: an anti-dualist account of interpretation, in *Objectivity, Relativism, and Truth, Philosophical Papers,* vol. 1, Cambridge University Press, Cambridge.

[6]Boorse, C. (1975) On the distinction between disease and illness. *Philos. Public Affairs* **5,** 49–68.

[7]Boorse, C. (1977) Health as a theoretical concept. *Philos. Sci.* **44,** 542–573.

[8]Temkin, O. (1963) The scientific approach to disease: specific entity and individual sickness, in *Scientific Change: Historical Studies in the Intellectual, Social and Technical Invention from Antiquity to the Present,* Crombie, A. C., ed., Heinemann, New York.

[9]Brown, M. W. (1985) On defining disease. *J. Med. Philos.* **10,** 311–328.

[10]Whitbeck, C. (1978) Four basic concepts of medical science, in *Proceedings of the 1978 Biennial Meeting of the Philosophy of Science Association,* vol. 1, Asquith, P. D. and Hocking, I., eds., Philosophy of Science Association, East Lansing, MI.

[11]Tartaglia, L. A., Dembski, M., Weng X., Nanhua, D., Culpepper, J., Devos, R., Richards, G., et al. (1995) Identification and expression cloning of a leptin receptor, OB–R. *Cell* **83,** 1263–1271.

[12]Galanti, G.-A. (1991) *Caring for Patients from Different Cultures,* University of Pennsylvania Press, Philadelphia.

[13]Johnson, M. (1993) *Moral Imagination: Implications of Cognitive Science for Ethics,* University of Chicago Press, Chicago.

[14]Gerr, F., Letz, R., and Landrigan, P. J. (1991) Upper-extremity musculoskeletal disorders of occupational origin. *Ann. Rev. Publ. Health* **12,** 543–566.

[15]Monaco, G. P. (1991) Moving ahead in cancer research: who pays for patient supportive care for participation in experimental/investigational trials? *Cancer Invest.* **9,** 85–92.

[16]Kraupl-Taylor, F. (1971) A logical analysis of the medico-psychological concepts of disease. *Psychol. Med.* **1,** 356–364.

[17]Katz, L. (1987) *Bad Acts and Guilty Minds,* University of Chicago Press, Chicago.

[18]Garrison, S. J. (1994) Geriatric stroke rehabilitation, in *Rehabilitation of the Aging and Elderly Patient,* Felsenthal, G., Garrison, S., and Steinberg, F., eds., Williams and Wilkens, Baltimore.

[19]Veldman, P. H. J. M. and Goris, J. A. (1995) Shoulder complaints in patients with reflex sympathetic dystrophy of the upper extremity. *Arch. Physical Med. Rehab.* **76,** 239–242.

[20]Wittgenstein, L. (1958) *Philosophical Investigations,* Anscomb, G. E. M. (trans.), MacMillan, New York.

[21]Pence, G. E. (1990) *Classic Cases in Medical Ethics,* McGraw-Hill, New York.

Introduction

Since it is conceptually possible for normal function to count as disease, malfunction cannot be a necessary condition for disease. Harm, not malfunction, is a necessary feature of the concept, making disease an evaluative idea. That disease is compatible with proper function has the consequence of removing disease from the category of the pathological, and it forces changes in the concept of a malady. Kaufman argues further that since the concept of harm can be objectively specified, disease determination is not relative, as normativists often assume. Disease can be understood as an aspect of "objective normativism." The normativist/neutralist debate about the concept of disease has been driven by false alternatives.

Disease

Definition and Objectivity

Frederik Kaufman

I

The concept of disease has been the subject of intense philosophical scrutiny, with much at stake. Is the notion of a disease purely descriptive, like the concept of triangularity, or is the concept of disease partly evaluative, like the concept of murder? Normativists argue that the concept is at least partly evaluative, whereas neutralists hold it to be purely descriptive.

Whether the concept of disease is purely descriptive or partly evaluative is significant because the nature of disease, and hence the foundations for medical intervention, are interpreted differently by each position. If the concept of disease is partly evaluative, then a condition cannot be a disease unless we disvalue it. Flies, for example, are pests because we do not like them; we do not dislike them because they are pests. Aside from our concerns, nature does not contain pests. According to normativists, the concept of a disease functions similarly—aside from our concerns, there are no diseases, only bodily conditions of one sort or another.[1] In their view, we call certain bodily conditions diseases (basically) because we do not like them. Neutralists, on the other

271

hand, deny the normativist claim about disease; they hold that the concept is purely descriptive. Rather than requiring an assessment of our interests, neutralists maintain that disease determination is an objective fact about the world.[2]

The issue of objectivity motivates much of the debate over the concept of disease. Since normativists think disease classification requires evaluation, they sometimes embrace what might be called "disease relativism." As the term suggests, since whether a bodily condition counts as a disease is contingent on our disvaluing it, and since values can apparently vary from culture to culture, so too can the conditions be taken as diseases.[3] Normativists are impressed by examples of diseases that are obvious projections of cultural values, such as the 19th century horror of masturbation, or the 18th century determination that slaves who sought freedom by running away suffered from a disease called "drapetomania."[4]

Disease relativism does not entail that different cultures will make wildly different disease determinations, however, since there is likely to be widespread agreement about certain basic values, such as the undesirability of suffering, lack of functional ability, and death.[5] The sharing of basic values explains, for the normativist, the broad convergence of thinking by diverse cultures on certain conditions as diseases, such as cancer, malaria, bubonic plague, beriberi, and the like. This convergence, for normativists, is no more remarkable than the fact that there is a somewhat similar convergence on the creatures taken to be pests, such as biting insects, rodents, and bugs that eat our food.

The broad convergence on the conditions deemed to be diseases allows normativists to be ojectivists of a sort about disease. Merely thinking a condition is a disease is not sufficient for it to be a disease, because how the condition links up to the shared values is not a matter of individual or cultural determination. That cancer is causally linked to suffering and death (things we disvalue) is not up to us. In this way disease determination need not be completely relative; disease determination can be as objective as our values, but widespread acceptance of the "fact/value" distinction means that normativists do not take themselves to be objectivists about disease in the same way that neutralists take themselves to be objectivists about disease.

Neutralists offer a different account for the objectivity of diseases. In their view, the general convergence on conditions taken to be diseases reflects the fact that they *are* diseases. Indeed, convergence on certain conditions being diseases is best explained by supposing that there is a matter of fact in virtue of which those conditions are diseases. The matter of fact often cited by neutralists on which our disease determinations converge is a species-specific function; making disease, in their view, a malfunction of one sort or another.[6] "Malfunction" is not evaluative; it indicates only a subnormal functional condition, empirically determined. Since a functional account of an engine can be given in purely descriptive terms, so too can its subnormal functions. Therefore, masturbation and drapetomania are not diseases, according to neutralists, because they do not involve species-specific malfunctions. Doctors in the 18th and 19th centuries were simply mistaken. Cancer and malaria are diseases because they do involve species-specific malfunctions. If for some reason we did not think these conditions were diseases, then we would be mistaken, because they are malfunctions irrespective of our thinking. Whether a human organism is functioning according to its evolutionary design is a perfectly objective, value-neutral, matter of fact on which our disease determinations converge over time.

II

Neutralists argue that malfunction is a necessary condition for disease. This conception of disease seems close to self-evident, and various historical explanations of disease, such as humor-balance or germ theory, rely on a notion of bodily malfunctioning. Malfunction is also what one typically finds built into dictionary and medical textbook definitions of disease. Despite its initial plausibility, however, malfunction is not likely to be a necessary condition for disease. If malfunction were a necessary condition for disease, then it would be conceptually impossible to have a condition that is a disease and not a malfunction. Yet it is at least conceivable for there to be a condition we would describe as a disease, but that does not involve malfunctioning. Suppose people past a certain age often became feverish, nauseated, and without medical intervention, tended to die. It makes sense to regard the condition as a disease. It also seems that this syndrome could be part of the species design. Perhaps the condition is a "self-destruct" mechanism with some sort of evolutionary advantage for the species.[7] On learning that the condition is part of our evolutionary design, perhaps we would deny that it is a disease, affirming instead that it is an aspect of our natural life cycle; but maybe not. Because of the effect this condition has on the individuals stricken, we might sensibly continue to regard the condition as a disease, and do what we can to combat it. If this is conceivable, then it is possible for proper functioning to count as disease.

That malfunction is not a logically necessary component of the concept of disease is evident from the fact that we could intelligibly regard the hypothetical condition described above as a disease independent of knowing whether or not the condition is a malfunction. The mere intelligibility of doing so establishes that malfunction cannot be a necessary condition for disease, because if neutralism were correct, it would be incoherent to think that a condition could be a disease *and* not a malfunction. If one were to insist, at this point, that proper functioning cannot be coherently conjoined with disease, because malfunction is logically necessary for disease, then one begs the question. Since it is at least coherent to think that a condition could count as a disease even although it is not a malfunction, neutralism cannot be correct.

The neutralist is making an implausibly rigid claim. If a condition is a disease, then, according to the neutralist, we know *a priori* that it must be malfunction, and if there is no malfunction, then a condition cannot be a disease. In the hypothetical case described above the neutralist must say that the condition is not a disease on discovering that it is part of the human biological plan.

Even if all actual diseases are malfunctions, this is a contingent fact with no bearing on the concept of disease, because we can conceive of a disease that involves normal functioning. The hypothetical disease described above might be approximated by cancer or heart disease. If a propensity for cancer or heart disease is, as some scientists speculate, genetically encoded, then these conditions may well be part of the human biological design. We might have been "designed" by evolution to succumb to these conditions to reduce the number of nonproductive older mouths to feed or to ensure that those in their reproductive years devote their attention to the young. Whatever the evolutionary explanation, if it is conceivable for cancer or heart disease to be part of the human biological design so that the individual's physiology is functioning properly when these self-destruct mechanisms operate at the appropriate time, then, again, malfunction cannot be a necessary condition for disease.

Certain conditions are diseases not because they are malfunctions—cancer or heart disease could be part of the proper-functioning of the species—but because they are bodily conditions that cause harm.[8] The concept of disease links up to the notion of harm, not malfunction, because there is no guarantee that "normal" biological function cannot harm. This is why it makes sense to regard certain conditions as diseases even if they turn out not to be malfunctions. They are diseases because of the harm they cause irrespective of whether they are malfunctions.

Proper biological function is compatible with harm. There are countless examples in nature of individuals who appear to be harmed for the evolutionary advantage of the species. Male animals frequently fight for reproductive access to females; it is not uncommon for the losers to be injured or driven off. The loser appears to be harmed. The male copulation response in certain insects is activated by decapitation by the female; this appears to harm the male.

One might object that functioning according to evolutionary plan ultimately cannot harm the individual, but this is an implausible *a priori* claim. One must argue that no matter what one's evolutionary design dishes out, necessarily, it is for the good of the individual. One would have to hold this for any conceivable evolutionary design. There is simply no reason to accept such a claim, and in the face of the many examples in which individuals at least appear to be harmed by their evolutionary design, it is sensible to think that in some cases they *are* harmed. Thus, the fact that a condition is part of normal evolutionarily selected biological functioning is no obstacle to its harming the individual, and thus is no obstacle to its being a disease.

Since harm is necessary for disease, and harm is an evaluative notion, the concept of disease is evaluative. Whether a bodily condition is part of normal functioning is beside the point; the main question is whether or not the bodily condition causes harm. However, determining that the concept of disease is evaluative does not straightforwardly establish normativism over neutralism, because, as I shall argue in the following section, harm might be an objectively descriptive fact.

Historically, the concept of disease may well have developed for those bodily harms in which the cause was obscure.[9] Animal bites, drowning, cuts, trauma, and so on, are not called diseases because the cause of harm is obvious, whereas bodily harms caused in ways that are not obvious need explanation. The causes of diabetes, dysentery, yellow fever, and the like, were certainly not obvious to early investigators, just as the causes of AIDS, cancer, Legionnaire's disease were (are) not obvious to modern investigators. The concept of a disease could have arisen to cover just these sorts of cases, because there is no theoretical difference between obvious causes of harm and obscure causes of harm. However, irrespective of whether the concept of disease arose in this way, the causal component of the concept does nothing to rule out harm caused by proper functioning as opposed to malfunctioning. A malfunction conception of disease overlooks this possibility, and thus cannot be an account of the concept. The conceptual compatibility of disease with proper function has a number of far-reaching consequences.

One important consequence is that we must reconsider the relation between disease and pathology—two basic medical categories. Disease is taken to be a subset of the pathological, by definition; hence "nonpathological disease" is a contradiction.[10] Pathology is understood with respect to the abnormal, a kind of departure from normal.[11] "Normal" is an idealized norm, not a simple statistical norm, since some pathologies, e.g., certain diseases, like colds or tooth decay, are statistically normal. The idealized norm is the functional design of a species as "intended" by evolution. Statistical norms will, of course, have to play a role in arriving at the idealized norm, but these and many other difficulties in arriving at the idealized norm for a species (or some subset, such as infants, males, females, and so on) need not distract us. When all is said and done, the idealized norm is a descriptive account of how the organism is "supposed" to function. Just as someone unfamiliar with internal combustion engines could, with enough study, figure out how one is supposed to work, so too the biologist can tell us how an organism is supposed to function.

Since it is conceptually possible for some diseases to be part of the species design, those diseases are not abnormal in the idealized sense; the diseased organism would be functioning as evolution "intended." Since the pathological requires abnormality, some diseases, therefore, cannot be pathological. Hence a major assumption about the relation between the categories of disease and pathology must be abandoned. The possibility of normal biological functioning counting as disease means that we must choose between not counting some diseases as pathological and allowing some normal functions to count as pathology, but each idea, namely, that disease is a kind of pathology, and that pathology requires abnormality, appears self-evident. Thus, resolution of this dilemma will require abandoning at least one self-evident conceptual truth. I propose that we give up the idea that disease is a kind of pathology, for the reasons that follow.

If physiology determines pathology,[12] then pathology is an evaluatively neutral biological concept; so disease, an evaluative concept, cannot belong to pathology at all. The concept of pathology is best used to describe certain deviations from how an organism is "intended" to work, as determined by evolution. Therefore, a hypothetical self-destruct mechanism selected for by evolution and part of the species design is not a pathological condition, even if it is a disease.

Disease belongs within the concept of a malady, as introduced by Culver and Gert[13] (but with modifications; *see* the following discussion). Maladies are the bodily conditions that harm, irrespective of whether the harm is "intended" by evolution. This means that we cannot infer harm from failure to function as designed (i.e., pathology), because, for example, failure of the hypothetical self-destruct mechanism is not a harm; whereas we can infer harm from disease, indeed, from any malady.

By "malady" Culver and Gert mean "roughly any condition in which there is something wrong with the person."[14] "Something wrong" is understood with respect to harm;[15] so maladies are those conditions that harm (or pose some risk of harm). Disease is one kind of malady; injuries, disabilities, dysfunctions, and so on are others. Whether a condition is a malady, however, is decided by what is normal for the species,[16] because what is normal for the species cannot constitute having something wrong with one. However, if what is normal cannot be a malady, and maladies are conditions that harm, then what is normal cannot harm. Yet this conclusion is precisely what must be rejected if we argue against a malfunction account of disease by observing that normal functions *can* harm.

My use of the notion of malady differs from that of Culver and Gert because determining whether or not a condition is a malady does *not* rely on the idealized norm. I shall use the term "malady" to refer to any bodily condition that causes harm to the individual, irrespective of whether that condition is normal (i.e., part of the proper functioning) for the species. The conceptual possibility of a normal function counting as a disease drives us to this conclusion. Determining whether one has a malady by determining whether one's condition is characteristic of the species overlooks the possibility of some maladies being characteristic of the species. Either Culver and Gert's account of malady is to be amended in the way I propose, or some diseases are not maladies.

I am proposing, in short, separating the concepts of pathology and malady. Pathology belongs to biology and is nonevaluative. Malady, the evaluative category containing the concept of disease, will thus supervene on purely biological descriptions, irrespective of whether the descriptions are pathologies.[17] Distinguishing the concepts of pathology and malady, locating disease within malady, and altering Culver and Gert's account of malady in the way indicated, resolves the dilemma produced by the conceptual possibility of normal functioning counting as disease.

Another consequence of realizing that malfunction is not necessary for disease is that a recent and highly regarded exploration of the nature of disease obviously fails.[18] After much careful work, Lawrie Reznek arrives at the following definition:

> A has a disease P if and only if P is an abnormal bodily/
> mental process that harms standard members of A's species
> in standard circumstances.[19]

But, again, if it is conceptually possible to have normal bodily processes count as diseases, abnormality cannot be a necessary condition for disease. It is surprising that Reznek builds abnormality into his definition of disease, because he argues *against* neutralists by pointing out that *normal* functions can count as diseases.[20]

III

The question of the objectivity of maladies, and hence of disease, remains. Since the concept of a malady is evaluative, it would seem that the kind of objectivity sought by neutralism is unobtainable, because evaluation is often thought to be subjective, in some broad sense. Perhaps widespread acceptance of certain values assures that there will be general agreement regarding which conditions are diseases, but there is a persistent worry that if fundamental values differ, disease (or malady) classifications can be irreconcilably different. All one can do in such a situation is oppose the attitude behind a particular classification, rather than adduce reasons against it.[21] Or so it seems.

Since harm is a necessary condition for disease (or malady), the objectivity of disease can be considered by way of the objectivity of harm. If whether someone has been harmed is fixed by certain nonevaluative facts, such that one cannot accept those nonevaluative facts while denying harm, then harm can be an objective determination.[22] Hence, the issue of disease objectivity could be pursued independent of contingent agreement regarding certain values, if harm can be objectively established.

Can harm be objectively established? Certainly not anything can count as harm. If a man claims that the color blue harms him, we want to know what he means. Are his eyes peculiarly sensitive to that color so that seeing it causes him pain; is blue a sign of disrespect, or does he think that the gods are angered by blue? Without a special background, it simply makes no sense to say one is harmed by the color blue. A special background is needed to connect blue to well-being, since harm must in some way detract from one's well-being. This, it might be thought, is progress only in the Pickwickian sense, because the idea of well-being is far from lucid.

Human well-being is an exceedingly complex idea, but surely aspects of it are objectively describable. If our well-being is tied in some very broad way to our nature as rational, self-aware beings, then our well-being will be capable of objective specification, if our nature is. Any plausible account of human well-being will have to give primacy to the exercise of our capacities for rational self-reflective existence. Exercising our capacities for deliberation, emotional development, interpersonal relations, freedom of action, political organization, physical achievement, and so forth, will, in very broad terms, constitute our good. We are harmed, then, when we are unable to exercise the capacities that constitute our good. Given that rational thought, freedom of action, physical achievement are constituents of our good, one cannot think that conditions that diminish or destroy those capacities are not harms, and it is a matter of fact whether or not a particular human capacity is impaired, because the normal range for these human capacities is empirically established. This means that at least some harms can be objectively determined.

Harm is frequently understood with respect to what negatively affects our interests. Interests are typically understood in terms of our rational or prudent desires. If so, the idea of interests then drops out as the middle term, making harm what negatively affects our rational or prudent desires; but I do not see how a plausible account of our rational or prudent desires could be given that was independent of our nature. Thus, I think the notion of human good, and hence harm, is best described with respect to our nature as reflective beings, rather than just in terms of our prudent desires. Therefore, the idea of harm cannot be specified apart from natural facts about us.

Harm is evaluative, but that is no obstacle to harm being a supervenient natural fact.[23] Harm would thus exist only at a certain level of description, but in this respect harm is no different than dogs, cats, tables, economies, and societies, which, although factual, also exist only at certain high levels of description. Disease, a kind of malady, which is a kind of harm, could similarly be a high level natural fact, since calling a condition a disease is a way of talking about harm. Moreover, that we make the disease determinations we do is plausibly explained by supposing that the harm is real, and disease determination involves detecting that fact.[24] In other words, we think certain conditions are diseases (i.e., harms) because they *are* harms. Surely this interpretation is intuitively more appealing than supposing that certain conditions are harms only because we think them to be. Our ability to detect harm need not be infallible, although it is likely to be reliable. This is why there is general agreement about the disease status of many conditions, but controversy about a few.

The normativist/neutralist dispute has apparently been driven by false alternatives: Neutralists think objectivity can be purchased only by a nonevaluative account of disease, and normativists think matters of fact cannot be evaluative. If there can be evaluative matters of fact, and harm can be such a fact, then it is possible for disease to be both an evaluative notion and an objective matter of fact. Rather than having to choose between subjective normativism and nonnormative objectivism, we can embrace objective normativism for disease.

Acknowledgments

I would like to thank my colleagues Richard Creel, Stephen Schwartz, and Robert Klee for helpful comments on earlier drafts of this paper.

Endnotes

[1]Sedgwick, P. (1981) puts the point clearly when he writes: "The blight that strikes at corn or at potatoes is a *human invention,* for if man wished to cultivate parasites (rather than potatoes or corn) there would be no "blight," but simply the necessary foddering of the parasite-crop. ... Outside the significances that man voluntarily attaches to certain conditions, *there are no illnesses or diseases in nature."* (p. 120, emphasis original) Quote from "Illness— Mental and Otherwise," reprinted in *Concepts of Health and*

Disease, Caplan, A. Engelhardt, H. T., McCarthney, J., eds., Addison-Wesley, Reading, MA. This is a classic collection on the topic of defining disease.

[2]Boorse, C. has vigorously defended neutralism in a series of influential articles; *see* (1975) On the distinction between disease and illness, *Philos. Public Affairs* **5,** 49–68; (1977) Health as a theoretical concept, *Philos. Sci.* **44,** 542–573; (1987) Concepts of health, in *Health Care Ethics,* Van de Veer, D. and Regan, T., eds., Temple University Press. *See also* Kendell, R. (1981) *The Concept of Disease and Its Implications for Psychiatry,* reprinted in Caplan, et al.

[3]See King, L. (1954) What is disease? *Philos. Sci.,* reprinted in Caplan et al., 1981, "Disease is the aggregate of those conditions which, judged by the prevailing culture, are deemed painful, or disabling, and which, at the same time, deviate from either the statistical norm or from some idealized status. Health, the opposite, is the state of well-being conforming to the ideals of the prevailing culture." p. 112.

[4]Engelhardt, H. T. The disease of masturbation: values and the concept of disease, reprinted in Caplan, et al., 1981.

[5]Goosens, W. (1980) Values, health, and medicine, *Philos. Sci.* **47,** 100–115. Goosens points out, "That a concept is normative does not entail value relativity, simply because the analysis may involve values which do not vary with persons or societies." p. 101.

[6]Boorse explicitly defends this view throughout his writings. Malfunction accounts of disease are typical in dictionaries and medical books.

[7]Versions of this argument can be found in Reznek (p. 129–133) and Goosens (p. 112). Other philosophers make this argument by implication. For example, Caplan, A. "The 'unnaturalness' of aging," in Caplan et al., 1981, argues that the naturalness, universality, and inevitableness of aging are not obstacles to aging's being a disease. Ruse, M. "Are homosexuals sick?", reprinted in the same volume, considers the possibility that homosexuality is part of human evolutionary design; *if* one argued on independent grounds that homosexuality is a disease nevertheless, then one would have to think that disease determination is compatible with species design.

[8]Reznek, p. 134; Goosens, p. 106. Incidently, appendicitis shows the neutralist position to be implausible. If the appendix has no function, as some speculate, then malfunction cannot be necessary for disease. Appendicitis is a disease irrespective of biological function because of the harm it causes.

[9]Kendall, R. 446 in Caplan et al., 1981; Reznek, p. 76; Hare, R. M., "Health," reprinted in his *Essays on Bioethics,* Oxford University Press, 1993, pp. 31–49, *see* p. 33.

[10]Hare (1993): "Not all conditions are pathological, but all diseases are, by definition." p. 32; Reznek: "There is a great variety of pathological or negative medical conditions. ... These conditions seem to fall into a number of mutually exclusive negative medical categories—not all of them are classified as diseases. Some are diseases —tuberculosis, syphilis, ... and lung cancer. Others are injuries—lacerations, gunshot wounds, burns ... " p. 65; "In summary, diseases (as opposed to other pathological conditions) do not form a distinct natural kind." p. 78; "I have argued that pathological conditions (of which diseases are one sort) can be understood in terms of conditions with a nature of a type that is abnormal and that leads to harm or malfunctioning." p. 99 and passim. That diseases constitute a subset of the pathological is simply an unquestioned conceptual truth.

[11]Boorse (1987), "The essence of the pathological is *statistically species-subnormal part-function"* p. 370 (emphasis original); Reznek, "If certain types of processes were normal, they would not be classified as pathological—it is not their nature being a certain type that makes them pathological, but their being abnormal that makes them so." Both neutralists (Boorse) and normativists (Reznek) accept the idea that pathology entails abnormality. This agreement is hardly surprising, since the point is definitional.

[12]Boorse (1987) writes: "My thesis is that whatever the correct analysis of biological function statements, *physiology determines pathology.* " p. 371 (emphasis original). Since Boorse regards diseases as a kind of pathology (p. 365), disease must be value-free. I am suggesting that we run things the other way: Since disease is value-laden, it cannot be a pathology.

[13]Culver, C. and Gert, B. (1982) *Philosophy in Medicine,* Oxford University Press.

[14]ibid, p. 66.

[15]ibid, "Briefly, to have something wrong with oneself is to have a condition ... such that one is suffering or has an increased probability of suffering some evil. ... [A]s we use the term "evil," it has no moral connotations; one could, without changing the meaning, substitute the term "harm" for "evil." p. 69.

[16]ibid, "When a person is suffering an evil, we decide whether he has a malady by determining if it is characteristic or members of the species to suffer a similar evil or increased risk thereof in this environment or circumstance." p. 74. "Thus, to be disabled [a kind of malady] is to lack some ability that is characteristic of the species at the appropriate level of maturation." p. 76; "thus 'increased risk' as the phrase is used in the definition of malady, must be understood as 'increased over what is characteristic of members of the species.' " p. 78.

[17]It is a mistake, then, simply to equate the concepts of pathology and malady, as Boorse does (1987); Note #15, p. 387.

[18]See the generally laudatory review by Pelligrino, E. (1989) in *Quart. Rev. Biol.* **64,1** (March), p. 101.

[19]Reznek, p. 162.

[20]Reznek, pp. 128–133.

[21]Reznek finds himself in this situation. He writes, "But, if judgments about the existence of norms are not true or false, and do not reflect any independently existing state of affairs, then it follows that judgments about disease are not true or false either. When we argue that we are correct to classify TB as a disease, and the South American tribe incorrect not to classify dyschromic spirochaetis [a skin infection the tribe finds attractive] as a disease, we are simply further endorsing the adoption of one norm rather than another." p. 168. Reznek does not take this to be *a reductio*. He endorses the idea that because disease determination is evaluative, it is noncognitive: "The fact that reference to some norm is essential gives us some reason to believe that judgments about disease are not true or false." p. 167.

[22]See Foot, P. (1978) "Moral arguments." reprinted in *Virtues and Vices,* University of California Press, pp. 96–109. She writes, "Anyone who uses moral terms at all ... must abide by the rules for their use, including rules about what counts as evidence for or against the

moral judgment concerned." p. 105; *see also* p. 106. Interestingly, in "Moral beliefs" (same book, pp. 110–131), Foot discusses the concept of injury; "We are most ready to speak of an injury where the function of a part of the body is to perform a characteristic operation" (p. 116). If Foot is proposing an analysis of the concept of injury, then more is needed, since we have not been injured if our (hypothetical) self-destruct mechanism fails to perform its characteristic function! Analysis of the concept of injury must include harm, not just failure to perform characteristic function.

[23]This point has been articulated by contemporary moral realists, such as Brink, D. (1989) *Moral Realism and the Foundations of Ethics,* Cambridge Studies in Philosophy; Railton, P. (1986) Moral realism, *Philos. Rev.* pp. 163–207; Miller, R. Ways of moral learning, *Philos. Rev.* vol. 94, no. 4 (October 1985), pp. 507–556.

[24]This argument is a generalized version of the argument by Sturgeon, N. in "Moral explanations," reprinted in *Essays on Moral Realism* (Sayre-McCord, G., ed., Cornell University Press, pp. 229–255. Sturgeon argues that moral explanations require positing moral facts; explanations involving evaluative terms should function similarly.

Introduction

The notion of value presupposed by the debate between Naturalists and Normativists in relation to the definition of disease is either impoverished or dualistic. A holistic model of human subjectivity is described in order to ground a richer notion of value and in order to suggest that the concept of disease encapsulates all levels of value generated by our subjectivity. This allows us to understand disease nonreductively and to explain why disease calls on us for a moral response.

Disease and Subjectivity

Stan van Hooft

There has been considerable debate in recent years surrounding the concept of disease. The question regarding what this concept refers to has been debated, some arguing that it is an entity that, in paradigm cases, invades the body, whereas others have argued that it is a holistic state of an organism involving internal disharmony or a dysfunctional mode of interaction with the biological and living environment.[1] As a result, there is an ontological (an invasion by a disease entity) and a physiological (an upsetting of internal balances) concept of disease. Writers have also disagreed on the classificatory criteria for diseases, asking whether they should be based on symptoms or causes, and whether the concept should embrace injuries and wounds, as well as other pathological conditions. Some have wondered whether the notion designates naturally occurring types of entity or event, or whether we have invented the concept for pragmatic purposes[2]. This has been called the problem of "taxonomic skepticism vs taxonomic realism."[3] There has been debate concerning whether the concept is a descriptive concept designating a dysfunctional state of the organism,[4] or an evaluative concept partly expressing attitudes to that state.[5] There has also been debate regarding the extension of the concept; whether it covers only physiological conditions or whether it is used, without change in meaning, with psychological conditions as well.[6] Are stuttering, hyperactivity, homosexuality, or drunkenness cases of disease? Further, there have been suggestions that disease is one of a set of concepts used for social labeling and that it can thus have repressive implications, as when Soviet dissidents were confined in mental asylums on the grounds that they suffered from mental disease.[7] The concept of disease raises the issue of the scope of medicine. Insofar as medicine deals with disease, so the definition of disease will set the boundaries of what is appropriate for medicine to concern itself with.[8] Underlying many of these debates has been a concern about the close connection perceived to exist between some concepts of disease and a dehumanising biomedical model of treatment.[10]

The dispute that I want to focus on in this paper is that between those who see the concept of disease as neutral or objective and those who see it as value-laden: the dispute between naturalism on the one hand, and normativism on the other. I support normativism, but, rather than adjudicate this dispute, I want to suggest that the meaning of the concept of disease, as it is used in the clinical setting of health care, is value-laden in a richer way than writers on either side of this debate have recognized.

I mention the clinical setting because this highlights the perspectives of those who care for the people with disease and of the patients themselves. This is important because the meaning of a term depends in part on the context in which it is used. Some of the debates about whether the concept of disease is descriptive or evaluative seem to suppose that this matter can be decided in the abstract. However, it may well be true that a medical researcher in a laboratory appropriately uses the notion of disease in a purely descriptive way, whereas a general practitioner is more aware of the evaluative connotations that the term carries when she uses it in her professional context. Again, in the debate between the ontological concept and the physiological, public health officials and pharmaceutical workers will use the ontological concept, whereas the physiological approach with its case histories and attention to the individual sickness will be appropriate to the clinical setting.[10]

What makes the caring context significant is that it has ethical implications for health care workers, who are frequently called on to make ethically sensitive decisions. The conventional approach of medical ethics and bioethics is to posit fundamental ethical principles or norms that can be used as the bases for deriving moral imperatives that will apply to the particular situation with which the health worker is confronted. As one author has put it,[11] ethicists take themselves to provide an algorithm by which decisions can be deduced syllogistically from uncontentious principles and norms. I have argued elsewhere,[12] that this is not a productive way of approaching morally complex situations and the apparent intractability of bioethical debates would seem to bear this out. My suggestion would be that what is required for ethical decision making is a sensitive awareness of what is at issue in the situation at hand. My ethical theory is particularist, virtue-based, and Aristotelian in its insistence on the importance of being aware of the moral demands that a situation contains. In the context of health care, morally difficult situations frequently involve disease. As such, sensitivity to the moral demands inherent in those situations will require a deep understanding of what disease is and what its significance is within human life.

Much of the literature agrees on the following points about the concept of disease. The first is that disease is involuntary. (We are victims of it. Drunkenness is not a disease, but alcoholism is.) Second, it renders us incapable of doing things we want to do and could otherwise do.[13] Third, it is unpleasant (although it can get me out of something worse, as when my asthma gets me out of military service). However, there are exceptions. Not all unpleasant debilitating experiences are diseases, e.g., childbirth, and it is also possible to be the victim of an asymptomatic disease that involves no unpleasantness because the victim is not aware of it (e.g., undiagnosed or asymptomatic HIV infection). So, in more formal terms, I would say that disease is harmful to its victims.[14] Fourth, disease is distinct from many congenital defects and from wounds and injuries. I will be following Lawrie Reznek in using the notion of disease to designate those pathological conditions of which the cause is not obvious, to differentiate them from wounds and injuries, which are pathological conditions of which the cause is patent.[15] Fifth, disease is a process with an onset, a typical course, and an outcome. Sixth, disease is not the contrary of health. There are other contraries of health, such as being unfit, which are not equivalent to disease. Last, notice that one should not assume that disease is by definition physiological since such an assumption would beg the question concerning whether mental illness is genuine disease.

Further, much of the literature on the concept of disease agrees that there is a structured set of distinctions to be drawn between the concepts of illness, disease, and the sick role.[16] In the context of such a set, the concept of disease is often seen as more objective and more closely related to the physiological. As one writer puts it, "Disease is something an organ has: illness is something a man has."[17] So, it is often suggested that disease is the pathological condition that causes negatively valued physiological or other changes. In this etiological form the concept seems to be both value-free and universal.[18] In contrast to this, illness is said to be the subjective state that responds to the changes caused by disease. Illness has a phenomenology that includes such states as pain, discomfort, and curtailment of the ability to do things.[19] Given this, it is clear that illness is a value term. It designates an unpleasant or painful state of partial or whole debility that people usually seek to avoid or seek relief from. Talcott Parsons has developed the notion of a social role that people who are affected by illness may fit into and that involves permission to not contribute to work and an obligation to seek cure and assistance. He called it the "sick role" with the result that cognates of the term "sick" are now most frequently used in this sociological sense.[20] Sociologists have also described "illness behavior" and the factors (apart from actual disease) that contribute to it.[21]

In this paper I will be arguing that in settings of patient care, the concept of disease does not belong at the lower, reductionist end of this three-level schema, but is evaluative in a very strong sense. In order to develop this argument, let us revisit some recent contributions to the debate on that issue.

Christopher Boorse has argued that disease is a purely descriptive concept. He has argued that disease is the theoretical concept while illnesses are "those diseases that have certain normative features reflected in the institutions of medical practice."[22] For Boorse, disease consists of a dysfunction afflicting a bodily process, whereas health is simply a case of each part of the organism fulfilling its natural function. Boorse offers the analogy of a Volkswagen that is in perfect working order when it fulfils its designer's specifications. To say this is not a value judgement because it may still not be a very good car compared to a Mercedes, for example. To say that it is in perfect working order is simply to describe a fact about it. In this way Boorse offers a purely descriptive analysis of health in terms of a well functioning organism, and disease is the equally value-free matter of the organism not being in perfect working order. Of course, the Volkswagen analogy has limitations. Science assumes that organisms do not have a designer in the way a car has, but science can discover the functions of our organs empirically (although not of the whole organism). As Boorse puts it, "A function in the biologist's sense is nothing but a standard causal contribution to a goal actually pursued by the organism."[23] Disease is a deviation from the natural functional organization of the species. Again, the notion of a deviation can be grounded on a nonevaluative description. Not all deviations are diseases, but "in general, deficiencies in the functional efficiency of the body are diseases when they are unnatural, and they may be unnatural either by being untypical or by being attributable mainly to the action of a hostile environment."[24]

Admittedly, the function of an organ or process can only be discerned in the context of an understanding of the wider goal-directed or teleological processes of the organism of which that organ or process is a part. After all, the disease entities may be functioning perfectly well and doing so in conjunction with relevant parts of the body. It is just that they are incipiently forming a different system and a different set of system interactions from that which would be functional within the host system. Once the operation and goals of the host organismic context is understood, functions can be attributed objectively and the failure of functions likewise described without appeal to values. It does not make sense, nor is it necessary, to say that the liver *wants* to purify blood, which is what we would have to say if we maintained that it was a frustration of value if it did not successfully purify the blood. For Boorse, it is simply objectively the case that the function of the liver within the organism is to purify the blood, and if it fails to do so, this is likewise an objective fact about it. It is diseased.

Boorse allows that disease is often undesirable, but this is a further and separate matter depending on a still larger context of human purposes and goals. That a person with a diseased liver will suffer frustration of her life goals does not imply that the dysfunctional state of the liver is inherently bad. It is made so by the further and noninherent fact that she has certain goals and desires. That it is undesirable for its victim is one of the reasons that we would describe the disease as an illness.[25]

Reznek, in contrast, endorses normativism. As he puts it: "Disease is to be understood in terms of the evaluative notion of being harmed. It also requires a reference to some norm—not all conditions that we would be better off without are diseases. The Normativist theory commits us to normativism—the thesis that the concept of disease is value-laden."[26] In the course of his discussion, he points out some difficulties with the notion of a function as defined with reference to goals, as it is by Boorse. Some things may contribute accidentally to my goals. Asthma may help me achieve my goal of getting out of the army, but it is not an instance of health. Again, some things, like chairs, have functions but are not part of any goal-directed system. Thus, there does not seem to be the logical link between functions and goals that Boorse requires. Reznek prefers an etiological theory that defines a function thus:

> The function of X (in Y) is Z if and only if (a) X does Z, and (b) X's doing Z makes a causal contribution to X's continued presence in Y in the right sort of way (by human intervention or via the mechanism of natural selection).[27]

On this account, the context of a goal given by a designer or by the process of evolution is needed to specify what the causal process is that explains X's occurrence. Reznek admits that the notion of dysfunction in this account will be value-free in that the goals and values of the organism are extrinsic to its functioning.

But this is no comfort for the naturalist because Reznek goes on to argue against the view that "being a process productive of a malfunction is a necessary condition for being a disease, and a sufficient condition for being pathological."[28] To make this point, he imagines a malevolent dictator who programs our genetic makeup so that we will self-destruct when we have outlived our usefulness. If this succeeded, then this process would fulfil the function it was given by its designer. It would not malfunction. Yet what the dictator had given us would still be regarded as a disease. Again, Reznek points to some processes which, in an evolutionary context, have no function (e.g., female orgasm, or the appendix) yet when they regularly fail to occur, or if they malfunction, we speak of disease. Indeed, from an evolutionary perspective, after reproduction has ceased, an organism has no further function, and so it cannot malfunction, yet it can have a disease, and so Reznek breaks the nexus between disease and dysfunction completely, tying the concept rather to the evaluative notion of harm, as we had noted earlier.

The notion of harm can only be explicated with reference to interests or goods, the frustration of which constitutes harm. As Reznek puts it, "someone is worse off, if and only if, his interests are impaired. But here someone's interests are impaired if and only if, his leading a good or worthwhile life is interfered with."[29] Yet Reznek admits that it is not easy to say what makes a life worthwhile except to say that it "*does* seem to have something to do with the satisfaction of desires, and the achievement of happiness."[30] On this account then, human goods or those values that can make a human life worthwhile provide the necessary context for understanding disease as a negative value. The disvalue that attaches to disease would seem not to be inherent in it but to arise from the fact that it frustrates the victim's pursuit of happiness and the goods of life.

Fulford argues differently from Reznek. For him both dysfunction and disease are value concepts.[31] He takes nonbiological functional objects to be the paradigm cases in which the meaning of the concept "function" is established. Such a nonbiological purposive context implies a designer such that the purpose of this designer is frustrated when there is dysfunction. This makes the value component of the notion of function depend on the purposes of the designer, and it also places limits on the valuations that may be imposed on a thing. It would be appropriate to say that this was not a good watch if it did not keep time, but not if it failed to serve as a paperweight. That a limited range of valuations are appropriate is argued for by Fulford when he suggests that purposefully designed things (and their parts) are capable of "functional doing."[32] Not everything that a watch can do—like be a paperweight—is what it functionally does, keeping time, along with the relevant movements of the mechanism, is, and this licences a move from "is" to "ought." If a thing is designed in a certain way for a certain purpose, then it *ought* to functionally do what it is designed to do. Dysfunction is a failure to do what it functionally ought to do and, as such, is a negative value.

I would add that this point also allows us to understand the notions of intrinsic value and extraneous value. It may happen to be true that a particular watch serves as a paperweight, but this would be an extraneous value. The watch's intrinsic value is that for which it was designed, it is the value that attaches to its doing what it functionally ought to do: namely, keep time, and it follows that it would only be an intrinsic disvalue if it fails to keep time.

How does all this apply to biological organisms and to persons? Fulford says that "biological functional objects—hands, hearts and so on—do not have designers. Yet they have "designed-for" purposes."[33] These purposes arise from biological needs such that an organism's fulfilling these needs in normal ways is a case of "functional doing" in the context of evolutionary teleology. It would follow that there would be intrinsic disvalue in any organismic dysfunction, but Fulford's crucial point is that human beings are capable of a richer form of functional doing, namely "intentional doing," in which the agents themselves set the purposes and goals that are being pursued. This form of doing can also be frustrated in a variety of ways. Where these frustrations are internal to the agent (so that they do not seem to have been done *to* one by someone or something else) and are not desired or under the agent's control, and where there is no apparent obstruction or opposition, we will have a case of illness.[34] Illness is related to intentional doing (or "ordinary doing" as Fulford describes simple actions) as dysfunction is related to functional doing. Illness is therefore a value concept just as dysfunction is. In illness, something fails to happen that ought to happen, namely, an agent's doing what he or she normally could do.

This also explains a difference in the use of the concepts of disease and illness. Only human beings and higher animals can be described as doing things intentionally.[35] As others have also noted, illness is more frequently attributed to human beings and higher animals, as opposed to disease, which can be attributed even to plants, because human beings and higher animals do things in a richer sense than plants (or watches).

A crucial point in Fulford's argument is that the structure of concepts that I described earlier, which makes disease the fundamental concept because it is the physiological concept, is in error. For Fulford, the fundamental concept is that of illness. The experience of not being able to do what one normally can do, and the concomitant experience of pain and discomfort, provide the basis for the concept of illness. Because it has the experience of frustration built into it, and because that which ought to happen fails to happen, this concept is clearly a valuational one. The concept

of disease, in contrast, designates a subcategory of illness distinguished from other cases of illness simply in being regarded as
disease and so directing attention to causes. The foundational
idea is that disease is causally implicated in "failure of action"
rather than that it is physiological. (Fulford is concerned to argue
this case because he wants to defend mental illness from the
skeptical effect of taking physical disease to be paradigmatic.
For Fulford any action failure can be thought of as disease, whether
its symptoms and etiology are physical or psychological.) Insofar
as disease is a subcategory of illness, it too, is a value concept.

But does this mean that the disvalue involved in the valuational concept of disease is simply that something fails to happen that ought to happen, as it is in the case of dysfunction? Yes,
but there is something more. The purposes that I pursue in my
everyday intentional actions will gain their importance for me
from the values that I hold. So values are the matrix of my actions,
and of my action failures resulting from illness. In this way, illness will be a disvalue to me that goes beyond its merely being a
frustration of my agency. It will also prevent me from attaining
the goals and values in pursuit of which I act.

This raises a question concerning whether the disvalue of
illness or disease is intrinsic to it. Even Boorse agrees that disease
is seen as a disvalue, but he denies that that disvalue is intrinsic.
Reznek attributes the disvalue of disease to its frustrating the
victim's good rather than to its being intrinsically dysfunctional.
Fulford can make good sense of the idea that something ought to
functionally do what it is designed to do and that it is therefore a
disvalue if it fails. This merely describes what ought to happen
given the teleological matrix in which the item functions. This is
not yet to say that when that happens which ought to happen, it
is a value pure and simple. A gun may fire perfectly well and in
that sense functionally do what it ought to do, but if it kills an
innocent victim, its doing so is not to be valued. What this shows
is that there is a value that can attach to functional doing but
that is distinct from its success in functionally doing just what it

ought to do. "Success" in this sense is an intrinsic value derived from the function the item is given by its designer or teleological context, whereas the further value is a value derived from a wider valuational framework in which the item is used and is, in my sense of the term, extraneous. Something's succeeding in functionally doing what it ought to do is not valuable in the same way that the consequences of that functional doing might be.

Human intentional actions are similar. I may go to do something and succeed in doing it. This success counts as a value in that my action has not been frustrated. But there are further values or disvalues that might attach to the action. A criminal who fires the gun at an innocent victim will judge (if he has reason to think about it) that that which ought to have happened has happened when he shoots, moreover, he will value this in the light of his own criminal goals. In a larger context of social values, however, others may judge that action negatively. In this context, an illness that prevents the criminal from firing the gun would be a value to society, even though it is a disvalue to the agent in that it frustrates his intentional action and prevents him from achieving his goal. So it seems that there is scope to go beyond the value of successful intentional action and find a further value in the action; a value that goes beyond the fact that the action came off successfully, and that derives from broader personal and social frameworks of evaluation. We can distinguish the disvalue of dysfunction or action failure from the broader disvalue that would be typically attributed to such failures, but might not be. This latter disvalue is clearly extraneous, and it might be this kind of broader value that Boorse is referring to when he says that the value component in the idea of illness or disease is an extraneous one arising from people's attitudes rather than an intrinsic one belonging to the concept of disease as such.

How do the notions of intrinsic and extraneous value, which I have analyzed in terms of "functional doing," apply to "intentional doing" abstracted from the broader social context? My discussion of Fulford shows that illness can be a disvalue in a number of ways. It is an intrinsic disvalue in that it involves action failure. The body will not do what it ought. Second, there is typically a further and less intrinsic disvalue in that the agent's goals are frustrated. (That this is less intrinsic is shown by the possibility that the agent might be relieved to find that an illness— a sudden paralysis, say—has prevented him from doing something that he had only decided unwillingly to do.) Third, there is the social disvalue or value of the action's not taking place. That this last can be an extraneous value so far as the agent is concerned is shown by the possibility that the agent, like the criminal, does not share those social values. In each case the valuation depends on differing teleological contexts. First, there is the inherent functional doing of the body. Second, there are the intentions of the agent. Third, there are the moral and cultural values of a society. Even if we leave aside the broader extraneous social values, there seems to be a question regarding the extent to which the agent's intended values are intrinsic or extraneous to the action. They are certainly not as intrinsic as is the value of the action occurring successfully.

Fulford needs an account of values that ties them to actions intrinsically, rather than extraneously, while being more than the value of the absence of action failure. For him, there is an analogy between functional doing and intentional doing. In my analysis, the value of a thing's successfully doing what it is designed to functionally do is intrinsic to it (and derives from the thing's designer). In the case of organisms, we assume no such designer but the analogy is still said to hold. However, the case of organisms who are human agents seems to introduce a new element. We give ourselves our goals in the way that a designer gives a functional item its goals. This would now seem to be a case of our imposing goals on our bodies understood as teleological systems that already have goals so that they are capable of functional doing. There seems to be a dualism here in which there are both functional values inherent in the body and intentional values arising from human agency.

The reconciliation of "is" and "ought" that Fulford espouses involves the notion of an "ought" that is less intrinsic to the agent than the "ought" that applies to the functioning of her body. If a car engine is designed to propel the car, then it is a true description of it to say that it ought to propel the car, but there is no sense in which the engine intends or "owns" this obligation as a part of its being. For organisms, including human organisms, the biological need to avoid harm grounds a prudential "ought." But this is intrinsic to the agent because it belongs to the body. For Fulford, this value is a function of the biological teleological framework in which the agent exists. The goals that an agent intentionally pursues seem not to belong to the agent in this intrinsic manner. If dualism is to be overcome then the intentional values of agents must be fully intrinsic to them also. I will argue that the values that frame intentional action are the kind to which agents can be committed and that are fully owned by them because they are not reducible to biological needs. These values constitute the good of the agent as that term is used by Reznek.

I would propose that the values in pursuit of which one acts intentionally are adopted or accepted so that they become intrinsic to the agent. It is not the case that agents ought to act in this way or that because this would fulfil some goal that is less intrinsic than the value of their own bodily functioning, but rather that agents find it important to be able to act in certain ways given the values that they hold. Only an intrinsic value of this kind is a value that can make a life worthwhile. Only such an intrinsic value could be described as a "good" for the agent. Only such a value would be so intrinsic to the agent that its frustration by illness or disease would be a disvalue in a richer sense than would be the failure of a bodily function. The notion of value that is needed to understand disease is therefore not a richer version of that of a goal or purpose as applied to functional objects or organisms. There are limits to the analogy between the functional doing of an object that is designed for a purpose and the intentional doing of an agent who pursues purposes.

I would argue that value in the rich sense that we need is a property of subjectivity as such. Therefore, in order to understand value and its relation to illness and disease, we now need to explore the notion of subjectivity.

Subjectivity is best understood by us (who are subjects) as exemplified by human beings. (I leave open the question concerning whether there are other beings who evince subjectivity. Human subjectivity is the only instance of subjectivity that we can know of directly through phenomenological reflection.) As Heidegger has put it, we are the being for whom being is an issue. If we read the first instance of the word "being" in this phrase as a noun and the second as a verb, we will glimpse what Heidegger is getting at. The fundamental, primordial, and inchoate impetus that undergirds human life is the drive toward being; toward "realizing" ourselves and our possibilities. Subjectivity is this impetus. The mode of being that we participate in as human beings is that of striving and seeking, of struggling and willing. We have an inchoate goal of being; of realising and distinguishing ourselves, of fulfilling our potential. I call this an inchoate goal because we do not become aware of it in this abstract and primordial form. We become aware of it in terms of the specific things we actually pursue in life, whether it be high academic grades, successful careers, loving relationships, respect of peers, health, or peaceful death. These are the particular goals and values, shaped by our culture and personal history, that we actually pursue, but our pursuing them is an expression of our primordial concern just to *be* in the way that a human being is. They are not values to be explained as arising just from evolutionary goals or social demands, or as if they were designed by the extraneous purposive frameworks in which we live.

Although it is internal to the person, it would be an error to suppose that this existential struggle to be was a purely mental or emotional phenomenon that did not involve our bodies. Indeed, Merleau-Ponty has famously argued that our bodies are the very expression of this quest.[36] For my part, I find it helpful to elaborate this dynamic picture of human subjectivity as comprising four levels of functioning.[37] The first of these is the level of the organism. At this level, the biological functions operate that are necessary for the continuance of life and the propagation of our species. We can gain an image of what this level would be like in its pure form by considering plants. The weed that breaks its way through the concrete of a carpark is evincing the sort of will to live that, in us, provides the basic level of our subjectivity. In us, this level comprises all the metabolic, visceral, and instinctive processes that keep us alive.[38] The existence of this level is what makes sociobiology relevant to, but not exhaustive of, a description of our subjectivity. The simplest value that arises at this first level of our subjectivity is the intrinsic value of survival. It would be misleading to call this a value in any complete sense since, at this level, subjectivity is blind to itself. At this level we evince little more than functional doing.

The second level is that of cognitive and affective reaction to the world around us. Lower animals display this level in its pure form when they respond instinctively to stimuli and display the kinds of affective stimulation that lead to appropriate action (such as hunger leading to feeding, or fear leading to flight). In the case of human beings, many of our preconscious immediate responses and desires will be either instinctive or learned as part of our acculturation. The most basic of these is our ability to recognize things for what they are and respond appropriately to them; an ability to which our having language is central. Our immediate aesthetic reactions, whether they be disgust at something repulsive, or delight in something beautiful, also belong at this level, as do the kinds of unconscious association that advertisers rely on to make their products seem attractive to us. Sartre's understanding of emotion as a way of colouring our experience of the world that is not in our full self-conscious control would be an example of a level two phenomenon in that it is a preconscious structuring of our world that is functional within our affective life. What traditional philosophy calls the transcendental level of human existence should be located at this level also. For example, Kant's transcendental ego is driven by a preconscious purposiveness that gives us useful knowledge, and the knowledge-constitutive interests that Habermas describes operate at this hidden level of our being.[39] Values that arise at this level include the intrinsic and inchoate values of joy at being alive and of desire-satisfaction, and the value that the world has for us insofar as it accords with our common-sense expectations.

The third level of subjectivity involves consciousness in the full self-aware form in which humans experience it. It is the level of functioning that involves thinking, planning, and deliberating about how we will meet our needs and fulfil our desires. Most of our everyday practical lives are focused on this level. It is the level that is most obvious and most present to consciousness on an everyday basis. It is the level at which we seek and find explanations for phenomena so that we can solve our practical problems, work, help those who are in need, and do the myriad of things that make up our daily lives. The notion of happiness as a fulfilment of our desires (as opposed to the notion of *eudaimonia* as a fulfilment of our whole being as agents) would be a third level concept.[40] The values that operate at the third level of our being are prudential values as well as those extraneous values, such as efficiency and wealth, that society might urge on us and that a given agent might adopt for prudential reasons.

The fourth level of subjectivity is the most distinctive and the most frequently neglected. It is the level at which our primordial need to find meaning in our lives is expressed in beliefs and commitments that are central to our integrity and our sense of ourselves. Our religious faiths, our moral principles, our search for pure knowledge, and our aesthetic values are just some of the cultural constructs that arise from, and structure, this level of our being. It is at this level that we express our commitments and at which we invest ourselves in that which gives significance to our lives—something that was most frequently done collectively through a shared culture in traditional societies, rather than individually as in modern societies such as ours. It is at this level that we can find our lives worthwhile and establish the values to which we find ourselves committed or to which we commit ourselves explicitly, and which it would jeopardise our integrity to flout. As such, these are values that are intrinsic to us, not in the sense that they represent our "designed-for" purposes, but in the new and richer sense that we fully "own" these values and invest ourselves in them.

Because it is our own integrity and sense of ourselves that is achieved by subjectivity at this level, there will be no psychological room for doubt and scepticism about the concepts and ideas that belong to this level. Concepts that operate at this level will be thought of as objective and as having an air of ultimacy attached to them. Plato's Theory of Forms was only the first in a long line of conceptualizations to posit objective, universal, and eternal realities as a counterpart to this level of our being. In the same way, believers in God think of Him as objectively real, and believers in morality think of its tenets as objectively binding.[41]

A full and rich human life will involve an integration and consonance of all four of these levels of our subjectivity so that whatever we do or experience will have some positive significance at all four levels. A simple act like eating will have organismic significance at level one, will satisfy desires at level two, will be the solution of a needs-based or scheduling problem at level three, and very often will have the significance of a social ritual that cements loves, friendships, or social relationships at level four. All these levels give eating its intrinsic value. Again, whereas the world will be a functional environment for us at level one of our being, and will appear intelligible to us at level two, and useful to us at level three, it appears as having importance and value at level four. The hidden and inchoate proto-values that arise from the lower levels of the model will provide the impetus and motivational power of the values that are thematised at higher levels.

Despite my talk of levels, this model of subjectivity is not hierarchical. There is no suggestion that the level at which the significance of our lives is established is more valuable or more important than the organic level. After all, it is clear that our functioning as an organism is a prerequisite for subjectivity at the other levels. We are not angels. Neither are we mere animals content with nothing more than the satisfaction of our desires, or *Homo economicus* or *Homo faber*, content with nothing more than the successful completion of our projects. We need meaning in our lives. The model is holistic because the meaningfulness of our lives depends on the integration of the four levels of subjectivity that I have identified. A person's values should be an integrated set arising from all four levels of subjectivity, and the good of a person is constituted by the fulfilment of her subjectivity at all four levels of her being.

This model of subjectivity can provide the basis for the notion of human good, and thus harm, that Reznek needs, for the teleological structure that both Boorse and Fulford require for the concept of disease, and for the notion of intrinsic values that, I have argued, is needed to understand disease adequately. This model overcomes the dualism suggested by attributing value to functional doing on the basis of a biological teleology and then to intentional doing on the basis of self-imposed goals. On this model, an agent's ideals, practical goals, comforts, and successful bodily functioning are all equally intrinsic values. Indeed, insofar as these values are interdependent, they constitute a single holistic intrinsic value. How might this help us understand the disvalue of disease?

The concept of disease and its cognate concepts might be mapped onto my model of subjectivity in a fairly straightforward way. If we regard disease as a bodily condition or invasion, then we would be apt to put it on level one. Organic dysfunctions occur at this level. It may be that our good at level one is our biological survival. Then again, it may be that, as individual organisms, we have no good of our own at level one. Perhaps at this level there is only evolutionary good. At this level, dysfunction may well just be failure to propagate one's genes. If this were true, dysfunction would be an objective notion in Boorse's sense, but it is an error to separate out one level of human subjectivity from the rest. Take Reznek's example of a physiologically grounded failure of orgasm. This is hardly likely to be experienced only as a biological dysfunction. Indeed, the harm that is suffered in this case has no organismic significance at all.

The notion of illness would seem to belong on level two. Illness is an affective state because it is an alteration in the way that the body operates such that its ability to act in certain ways is frustrated. This is experienced only after it has had an impact on how the organism reacts to, and moves within, its world. The inability to do certain things that marks illness in Fulford's account is a preconscious state. Although we, as conscious beings, can be aware of it and can respond in deliberate ways to it at level three, our state of well-being is already affected and our feelings and emotions are already elicited. In illness we present ourselves to ourselves in a different way. Whereas in health our awareness flows through our body, as it were, without hindrance and on to the things and people in the world with which we are practically concerned,[42] in illness, the flow of our agency is interrupted and hindered, if not stopped altogether, by our bodily condition.[43] This produces immediate reactions which structure our state of being. Among these will be suffering, discomfort, and the pursuit of compensatory means of acting. Pain will also be a direct bodily response to lesions of many kinds that structures our affective being. For self-conscious beings, such as ourselves, we will be aware of many of these responses as a complex of feelings and emotions.

At the third level of our being we will respond to this suffering by deciding what to do. We may ignore it, we may self-medicate or rest if we think the problem is not too serious, or we may seek medical help. With any of these responses we enter onto the third level of our being, at which our illness has become a practical problem for us such that we engage in illness behavior. In the broader social context, this is the level at which we make provision for health care, establish hospitals and clinics, and develop the professions of medicine, nursing, and other forms of therapeutic practice. It is at this level, too, that Talcott Parsons notion of the sick role belongs, since this is defined as a functional social role that emerges as a practical response to the disablement for everyday tasks that illness (or injury) brings.

Among the prudential values that operate at this third level of our subjectivity are those relating to health and illness. As Margolis puts it, "human beings, viewed in a sense that is relatively neutral to their condition as animals or persons, subscribe to a characteristic set of (what may be called) prudential values—avoidance of death, prolongation of life, restriction of pain, gratification of desires, insuring security of person and body and property and associates, and the like."[44] Medicine is a means for managing some of these prudential interests that belong at the third level of our being, and this level is where I locate what I have called extraneous values. There may be a prudential reason (like avoiding military service) for valuing being in a sick role, but this does not mean that the illness is intrinsically valuable. In this example, the possibility of differing extraneous prudential evaluations pertains to the sick role, not to the presence of disease. Nevertheless, this distinction between prudential evaluation and intrinsic worth is what allows us to explain why Boorse could say that the concept of disease is an intrinsically value-free one. Taken just as a level three phenomenon, disease and its concequences might attract differing kinds of prudential evaluation in differing circumstances. However, this says nothing about the intrinsic value or disvalue of disease for a human subject taken as a whole.

But, whereas it may seem illuminating to map disease, illness, and the sick role onto the first three levels of my model of subjectivity in this way, does this mapping do justice to the concept of disease and to the phenomenon of human subjectivity? It seems to leave us with two problems. First, because it makes the concept of disease seem fundamental (insofar as our bodily existence is fundamental to our subjectivity), it is at variance with Fulford's point that illness is the more fundamental concept in the set. Second, it seems to leave no role for the fourth level of our being, but it is precisely this level that secures the intrinsic nature of the values by which we live. Without this level we would have intrinsic values at levels one and two and extraneous values at level three, but this would leave a human subject without an integration of values and without a sense that life was worthwhile. It follows that the notion of disease needs to obtain its negative value through its connection with my holistic notion of the human good.[45] In order for its disvalue to be intrinsic to it it needs to be analyzed in a way that includes the fourth level of our subjectivity.

I would suggest, therefore, a new way of mapping the relevant concepts onto my model of subjectivity. I propose that the concept that belongs with the organic level of our subjectivity is that of "pathological condition." This concept will include those of injury and wound as well as disease, but this need not concern us. The key point about this concept is that it designates a condition of the body that is counter to the proper functioning of the body. It is, in short, the concept of dysfunction on which Boorse places so much stress. If Boorse thinks that this concept is a value-free one, it is simply because it is an impoverished one. It relates to only one level of human subjectivity. Yet, against Boorse, I would now say that, although it may be a value-free concept when applied to plants, it is not when applied to human beings, since human beings have intrinsic goals by virtue of their subjectivity, even at the organismic level. Incidentally, another way in which the concept of dysfunction is impoverished in the context of caring for patients is that it is typically used to refer to bodily parts. It is the heart, or the liver, or a limb that is said to be in a pathological condition rather than the body as a whole. In scientific settings the word "disease" may be appropriately used for pathological conditions or agents, but in clinical settings the word is not rich enough.

I would leave illness and the sick role on levels two and three, respectively, to explain what Fulford is saying when he argues that "illness" is the more fundamental concept of the set. There is something important going on when our immediate reactions are deformed and curtailed by a pathological condition such that our ability to do what we normally can do is frustrated. The pathological condition issues in an affective reaction that flows onto the other levels of our being such that, at the third level, we display the reactions of a sick person as defined within our culture if the illness is of an appropriate kind (sufficiently serious or debilitating or unusual or threatening to others). We refer to the whole organism as suffering an illness, rather than just an organ within the organism. Indeed, for Fulford, it is important that the notion of illness is applied to persons rather than just to organisms, because for him, the concept is used to report the action failure that a person is aware of themselves as experiencing, and in this way, the notion of illness spills over onto the third, self-conscious and pragmatic level of our subjectivity.

I would want to suggest that, rather than belonging on level one of my model, the concept of disease takes a distinctive connotation from the fourth level of our being. Disease is typically thought of as an evil. What I mean by this is that it is an experience or presence (either actual or potential) in a person's life that that person needs to assimilate into a structure of meaning and that is understood by our culture, not just as a harm, but as an evil. Most cultures think of disease as an attack on the bodily or personal well-being of a person that is to be feared and hated, that approaches us mysteriously, and that has the potential to cause suffering and death. It can come from outside of the person, or it can originate within. It can be a curse imposed by enemies, a punishment from the gods,[46] or it can be a deliverance of fate, but in each of these conceptions, the meaning that is given to disease derives from a faith or a cosmology that is the form that the ultimate commitments of subjectivity are given in those cultures.

A clue to what is being said here is provided by Reznek, who argues that we have retained the disease concept, despite a reductionist tendency driven by science that would favor the concept of a pathological condition, because of an ancient tradition in which disease was associated with mystery. A disease—as opposed to an injury—was something the cause of which was not obvious.[47] As such it was apt to be understood holistically rather than physiologically in terms of strange and malevolent powers beyond our knowledge. Disease (unlike injuries, which were treated as bodily problems to be solved with bandages and the like) was an event that called for the intervention of priests, shamans, or witch doctors.[48] For me, this signals that we are talking about a level four event: an event or process that is to be structured by subjectivity as part of the complex of ideas that expresses our pursuit of meaning.

As reported by medical anthropologist G. Lewis,[49] the Gnau people of Papua New Guinea think of disease not as instances of a general type of ailment that might strike anyone, but as personal, unique, and individual events that occur to them in response to some event that is a part of their own personal history. He describes a case of a person with a bad headache who attributed this to the cutting down of a tree that had significance in the kin structure of his tribe. In another case, a woman fell ill (without there seeming to be a physiological cause) and attributed the illness to arguing with relatives. In cases such as these, not only the significance of disease but its very identification is tied to the meaning that the episode has in the context of the victim's life and relationships.

However, not all meaning-conferring constructions of disease refer to taboos, magic, supernatural agencies, or powers. "Chinese medicine views illness as an expression of the personal violation of a person's own nature and calls on the person to become aware of how he or she is interfering with the flow of nature, both within and without."[50]

It might be objected that such examples are not relevant to our modern Western understanding of the notion of disease. For us, science has displaced such superstitions. But it should not be thought that modern scientific thinking has countered the tendency to think of disease in these ways. After all, we are subjects in the same way that preliterate people were subjects and we have the same need to give meaning to our lives and to the things that happen to us. Although a great many people no longer believe in curses, divine punishments, or the benevolent influence of providence (although there are still many who do), the question "why me" still resounds in wards and sick rooms. This question, whether or not it receives a coherent answer in these secular times, expresses the need that subjectivity has to find meaning in what happens to it. Disease is still experienced by most people as an undeserved affliction challenging their sense of justice,[51] and justice is a level four concept. In addition, it seems to be a threat to our faith in the progress of science in that it is a personal disaster that modernity has not been able to prevent, and our faith in science belongs on the fourth level of our subjectivity. Moreover, even in modern English, the word "disease" continues to carry connotations of rottenness and corruption, which are concepts with a deeper resonance than is found in the practicalities of everyday life.

It follows that the concept of disease is a bearer of meaning that is richer than that of a bodily pathological condition, illness as reaction to action failure, or the sick role as practical social construct. It is the stuff of metaphor and myth. The essays of Susan Sontag on cancer and AIDS bear witness to this.[52] She argues that diseases of these kinds (for which, once again, significantly, the causes are not well understood) play a cultural role continuous with the notions of demonic possession and sin as well as those of enobling affliction and victimhood. These are ways of structuring the experience of disease (especially fatal and wasting disease) to give it meaning. As Sontag puts it, "TB was a disease in the service of a romantic view of the world. Cancer is now in the service of a simplistic view of the world that can turn paranoid. ... For the more sophisticated, cancer signifies the rebellion of the injured ecosphere: Nature taking revenge on the wicked technocratic world."[53] Leaving aside the details of her thesis, it is clear that this quote is redolent with the sort of language that bespeaks the fourth level of our subjectivity. Crucially, what Sontag highlights, in contrast to understanding diseases in terms of their specific physiological causes (which would be level one thinking), is an attempt to attribute blame for their occurrence in the lifestyle of the victim, the levels of daily stress imposed by social arrangements, or in the environmental conditions that obtain. But this attempt to apportion blame is an attempt to give meaning to the disease. It is an attempt to answer the question, "Why me?" and to place it into a context of cosmic justice, and this means that disease is a disvalue that belongs at the fourth level of our subjectivity. It is not just that our third level goals are frustrated by action failure, but also that a threatening, disempowering, and evil fourth level reality has entered into our lives.

At the fourth level of our being, health and disease are contraries. Health is a good that we all strive for as an ideal and attribute to ourselves as part of our meaning-conferring construct of self-identity (even, sometimes, when we are suffering disability or disease)[54], whereas disease is an evil that we need to accommodate ourselves to while preserving a sense of the worthwhileness of our lives. The ideal of health can be manifest as a cult of vitality celebrated in art and sport, whereas the sense of evil that attaches to disease is manifest in the preventative significance given to practices of hygiene and in the isolation of its victims in sterile institutions. The representation of disease in art and literature that Sontag and others have explored provides evidence of this. There is very little poetry in pathological conditions, but disease can be a subject for art because it is a concept to which connotations of human significance readily attach. The notions of defilement, corruption, deformity, and suffering are notions with an aura of significance that more clinical categories do not have. The fact that our relation to disease is normally one of fear and loathing does not negate this point. After all, we define ourselves not only by what we love and commit ourselves to, but also by what we contrast ourselves with as our "Other." At the fourth level of our being, disease is part of that which is alien to us and defines our being by difference from it. Health is that which we absorb into our ideal being, whereas disease is that which is foreign to our ideal being. This explains the significance of both Reznek's point that the concept is retained in order to designate illnesses with mysterious causes and also the lasting attraction of an onto-logical concept of disease as an invader of the body.

What is needed to apply these insights to our own modern situation is an exploration of the kind of myths and metaphors that are available in a multicultural secular society, such as ours, to enable people to make sense of their diseased condition. Such an exploration will be empirical in the way that sociology and anthropology are empirical. That is, it will require empathy with, and *verstehen* of, the meaning-giving life beliefs of others. It is likely that such an exploration will disclose that, for many people, the concept of disease will take on meaning-giving power from the religious faiths that those people have. The notions of provi-dence and the will of a benevolent God will provide many victims with the basis for maintaining the meaning and worth of their lives. For others, in a modern "disenchanted" culture, such as ours, the meaning of disease will arise from its association with science. Insofar as the concept is used in the context of the bio-medical model it will acquire its aura and its connotations from that model. The germ theory of disease, amplified in terms of bacteria and viruses, and the notion of medical science as the designer of magic bullets to destroy these invaders can become an icon in the faith of modern secular humanity.

Ironically, understanding disease as a bodily dysfunction on analogy with a dysfunctional machine is precisely the kind of metaphor that would be needed to understand disease in secular terms, rather than as a punishment, a curse, or an expression of the will of God. Perhaps the impetus behind Boorse's insistence that the concept of disease is value-free is that only such a concept escapes the false comforts of a mythical conception. A thoroughly modern secularism loves fate. It believes that things can go wrong; that material processes are subject to failure and break-down. This love of fate would be a Stoic determination to go with the flow of nature and to accept whatever it delivers.[55] This meaning may be austere, but it can be comforting in the way that a sense of tragedy is comforting. Like bushfires, earthquakes, the sinking of ships, and other accidents, disease is just something that happens, and it is to be accepted honestly and bravely when it does happen. Attributing it to nonnatural powers would be, for this attitude, a form of escapism, but even if this were what Boorse was getting at, it would still be the case that the concept of disease that carries this connotation of accepted tragedy would be a construct of subjectivity designed to give meaning and worth to the life of the victim. Along with the several religious frameworks that are available in our culture for giving such tragedies as disease a meaning in human life, there is the secular faith that disease is just something that happens and that we must make the best of. Like any faith, this allows us to affirm the event, to own it, and to not blame anyone or anything else for it. On this view there is no answer to the question "Why me?" Accepting this is also a meaning-giving faith expressing our subjectivity at the fourth level of our being.

It might be objected that my thesis seems to leave the concept of disease hopelessly subjective and relative to mythical and metaphorical construction (a problem for the defenders of mental illness as not being a myth). Health workers need an objective concept. The burden of Susan Sontag's essays was to show the harmfulness of the metaphors that are applied to such diseases as cancer, syphilis, AIDS, and others; for example, in the way the metaphors of plague and scourge allow blame to be attached to the victims.

Now it is true that in diagnostic and treatment settings, disease concepts will assimilate themselves to descriptions of bodily pathological conditions to have the kind of objectivity needed for scientific practice, whereas at the fourth level the connotations of the concept of disease might seem subjective and relativistic. However, I have said that the faith and commitment concepts that subjectivity develops will seem objective to it. Disease will be understood as objective even at the level of meaning. The believer in myth sees himself really affected by the curse of another, and the HIV-positive person might see himself as a real victim of a plague. Disease may be a myth, but that will be a form given to its experienced reality.

The objector will not be satisfied with this answer since it merely restates the problem in more acute terms. A further answer that I could give is that no matter how much the cultural constructs that shape the meaning of disease are challenged and undermined, subjectivity will seek some new construct. The key point is that disease is always going to be the sort of thing to which metaphorical and mythical meanings can attach themselves. This shows that it is a fourth level phenomenon. The meanings that it is given can be altered, and many of them ought to be, but the crucial point is that some meaning or other must be given to it since disease is not experienced just as a pathological condition, as illness, or as the sick role.

The implications of this for health care are clear. A health care worker when confronted with a diseased patient is not confronted just with a body with a pathological condition, or a being whose reactions are curtailed, or a person whose social and practical life is in turmoil. These descriptions are all true of the patient, but they are not complete. The crucial point is that the patient is one the meaning of whose life has changed. The patient's subjectivity is not just frustrated and redirected into a new social role, that of being a patient. The patient's subjectivity suffers a diseased embodiment. Her life must and will achieve a new integration by the application of a disease concept that will give meaning to her new existence. It is not true that this will be a purely descriptive concept. It is true but shallow to say that this will be a valuational concept. It is a concept of deep value or, in my sense, a concept that belongs to the fourth level of our subjectivity; a concept that shapes our commitment to life itself. Even without knowledge of what patients might believe in particular cases, the implication is clear that health care workers need to understand the concept of disease as a form available to give expression to the subjectivity of patients when under threat from something alien, rather than the name of a class of pathological conditions or ill-nesses of which the only implication is the nature of the prognosis and treatment that should be given.

That the concept of disease is a value-laden concept consti-tuted by the fourth level of our subjectivity has immediate moral implications for health care workers (and for others, such as the family of the patient). A diseased patient is one who calls on our moral response in a direct way, through the shared understanding of the significance of his or her situation (which differs from that of an injured patient). It is not enough to say with Caplan, "Choos-ing to call a set of phenomena a disease involves a commitment to medical intervention, the assignment of the sick role, and the enlistment in action of health professionals."[56] There is more: namely, the acknowledgement that the patient now has or needs a new structure of meaning for his or her life, at least for the time being. The sick role by itself does not provide this meaning because, through the experience of action failure and loss of control over one's own life, it militates against finding one's life worthwhile.

What is the form of the ethical response that health profes-sionals are called on to make? I have already indicated in the first section that sensitivity is the first requirement. Knowing that disease is not just a bodily pathological condition, but a new structure of meaning in the life of the patient, is the key to avoid-ing the dehumanizing effect often attributed to the routinized application of the biomedical model. Treating the patient as a person involves being aware of all four levels of subjectivity that constitutes that personhood. As well, health care workers need to acknowledge the role that the concept of disease plays in their own lives. The notion of professional commitment includes that of an understanding of the central ideals and values in a professional life. Insofar as health care work involves a commitment to the promotion and restoration of health and to the eradication of disease and relief of its symptoms, health care workers ought to have a deep understanding of the values involved. This will ground not only their caring for patients, but their own sense of worth and dedication to the everyday tasks that belong to their profession.

Acknowledgment

This paper was prepared with the assistance of a research grant from the Faculty of Arts at Deakin University and with the research assistance of Peter Rzechorzek.

Notes

[1]*See* Dubos, R. (1959) *Mirage of Health: Utopias, Progress and Biological Change,* Harper and Row, New York, Chapter V for an interesting discussion of the history of this debate. *See also* Chapter 1 of Cassell, E. J. (1991) *The Nature of Suffering and the Goals of Medicine,* Oxford University Press, New York.

[2]King, L. S. (1981) What is disease, in *Concepts of Health and Disease: Interdisciplinary Perspectives,* Caplan, A. L., Engelhardt, H. T. Jr., McCartney, J. J., eds., Addison-Wesley, Reading, MA, pp. 107–118 argues that disease is a classification that alters with knowledge and interest.

[3]Reznek, L. (1987) *The Nature of Disease,* Routledge & Kegan Paul, London, p. 25.

[4]Boorse, C., On the distinction between disease and illness, in Caplan et al., op. cit., pp. 545–560. For Boorse, disease is a deviation from the natural functional organization of the species. "In general, deficiencies in the functional efficiency of the body are diseases when they are unnatural, and they may be unnatural either by being untypical or by being attributable mainly to the action of a hostile environment." pp. 552,553.

[5]Margolis, J., The concept of disease, in Caplan et al., op. cit., pp. 561–577. *See also* Reznek, op. cit., and Fulford, K. W. M. (1989) *Moral*

Theory and Medical Practice, Cambridge University Press, Cambridge.

[6]Flew, A. (1973) *Crime or Disease?* Oxford, New York.

[7]Michel Foucault develops this idea in *The Birth of the Clinic: An Archaeology of Medical Perception* (1973) (Sheridan Smith, A. M., trans.), Tavistock Publications, London.

[8]Gillon, R. (1986) On sickness and on health, *Br. Med. J.* **292,** 1 Feb., pp. 318–320.

[9]Engel, G. L., The need for a new medical model: a challenge for biomedicine, in Caplan et al., op. cit., pp. 589–607.

[10]Temkin, O., The scientific approach to disease: specific entity and individual sickness in Caplan et al., op. cit., pp. 247–263, p 262.

[11]MacLean, A. (1993) *The Elimination of Morality: Reflections on Utilitarianism and Bioethics,* Routledge, London and New York.

[12]*See* this author's (1995) *Caring: An Essay in the Philosophy of Ethics,* University Press of Colorado, Niwot, Colorado.

[13]This is the key to Fulford's analysis, op. cit.

[14]This is the key to Reznek's analysis, op. cit. Even here there can be exceptions in that a person may become morally better as a result of suffering a disease.

[15]Historically "there was a need to explain suffering in the absence of an obvious cause." Reznek, op cit. p. 77. Even now, Reznek argues, when we do understand the causes of many diseases, we continue to preserve the distinction between diseases and other pathological conditions. The distinction between medicine and surgery would have preserved the distinction also.

[16]This three way classification is also accepted by Radley, A. (1994) in his *Making Sense of Illness: The Social Psychology of Health and Disease,* Sage Publications, London. *See also* Helman, C. G. (1981) Disease vs illness in general practice, *J. Royal College of Gen. Practitioners* **31** (September), 548–552, and Barondess, J. A. (1979) Disease and illness—a crucial distinction, *Am. J. Med.* **66** (March), 375,376.

[17]Helman, C., op. cit., quoted in Calnan, M. (1987) *Health and Illness: The Lay Perspective,* Tavistock, London, p. 134.

[18]As one writer puts it, "It should be clear that a consequence of applying or using this essentially biologistic framework regarding disease

is that specific diseases can then be said to be *universal* or
transcultural occurrences." Fabrega, H. Jr., Concepts of disease:
logical features and social implications, in Caplan et al., op. cit.,
pp. 493–522, p. 497.

[19]Sedgwick, P., Illness—mental and otherwise, in Caplan et al., op. cit.,
pp. 119–129, and Fulford, op. cit.

[20]Parsons, T., Definitions of health and illness in the light of american
values and social structures, in Caplan et al., op. cit., pp. 57–81.
See also Veatch, R. M., The medical model: its nature & problems
in Caplan et al., op. cit., pp. 523–544, who says: "To be sick is to
have aberrant characteristics of a certain sort which society as a
whole evaluates as being bad and for which that society assigns
the sick role." p. 526. Various other concepts are linked to this
three-part distinction, although not always systematically. Thus,
we describe a physiological pathological condition as a disease if
it is a process but as an impairment if it is a static and persistent
condition, such as a malformed limb. Again, the notion of illness
as a description of a subjective experience is confined to
experienced processes, whereas if the experience is of a state
which is static and persistent, it might be referred to as a disability.
In a similar way sickness as a social role that results from a
pathological process might have its static and persistent counterpart
in the social role of being handicapped (Susser, M., "Ethical
components in the definition of health," in Caplan et al., op. cit.,
pp. 93–105). Given that these latter terms have a degree of
unacceptable social stigma attached to them, however, they are
frequently changed in order to avoid the negative connotations
that come to apply to them. So, disability and being handicapped,
are now sometimes referred to as "being challenged" or "being
differently abled." Similar attempts to overcome the stigma
attached to the sick role are represented by the tendency to replace
the word "patient" with that of "client"; or even "health care
services consumer."

[21]Mechanic, D., The concept of illness behavior in Caplan et al., op. cit.,
pp. 485–492.

[22]Boorse, op. cit. p. 550.

[23]Ibid, p. 551.

[24]Ibid, pp. 552,553.

[25]In a footnote, Boorse says he ceded too much to normativism in saying that illness is disease laden with values. "Illness is better analyzed simply as systemically incapacitating disease, hence as no more normative than disease itself." Ibid, p. 560.

[26]Reznek, op. cit., p. 170.

[27]Ibid, pp. 116,117 (my formulation combines several points that Reznek makes).

[28]Ibid, p. 126.

[29]Ibid, p. 150

[30]Ibid, p. 151.

[31]Fulford, op. cit., pp. 36–56.

[32]Ibid, p. 93.

[33]Ibid, p. 107.

[34]Ibid, p. 119.

[35]Ibid, p. 123.

[36]Merleau-Ponty, M. (1962) *The Phenomenology of Perception* (Smith, C., trans.), Routledge & Kegan Paul, London.

[37]The model of subjectivity that follows is elaborated fully in van Hooft, op. cit., Chapter 3.

[38]That these processes are both hidden and central to our lives as conscious beings has been argued in by Leder, D. (1990) in his, *The Absent Body,* The University of Chicago, Chicago, Chapter 2.

[39]Habermas, J. (1972) *Knowledge and Human Interests* (Shapiro, J. J., trans.), Heinemann, London. Habermas identifies a technological interest in the control of nature, a communicative interest in mutual understanding, and an emancipatory interest in overcoming material and social forces that prevent us fulfilling our being. These interests are not present to consciousness although they operate to constitute knowledge.

[40]From which it would follow that utilitarianism is a third level moral theory and therefore inadequate to the fullness of human existence.

[41]*See* Kant, I., *Religion Within the Limits of Reason Alone* for a rigorous example of this kind of thinking.

[42]The way in which the body "disappears" in ordinary life and re-emerges as a problem in illness is described by Leder, D. (1990) in *The Absent Body,* The University of Chicago, Chicago.

[43]A graphic description of this is given by Toombs, K. (1992) The body in multiple sclerosis: a patient's perspective, in Leder, D., ed.,

The Body in Medical Thought and Practice, Kluwer Academic, Dordrecht, pp. 127–137.

[44]Margolis, J., The concept of disease in Caplan et al., op. cit., pp. 561–577, p. 574.

[45]As Reznek puts it, "I have argued that causing a biological malfunction is neither necessary nor sufficient for being a disease (or pathological). It is not necessary because interference with some nonfunctional systems count as diseases. It is not sufficient because interference with self-destruct systems with functions do not count as diseases. What explains why these are counterexamples is that these processes either have or lack a connection to harm." op. cit., p. 155.

[46]Kopelman, L. (1988) in The punishment concept of disease in Pierce, C. and VanDeVeer, D., eds., *Aids: Ethics and Public Policy,* Wadsworth, Belmont, CA, pp. 49–55, gives a modern, secular critique of such views.

[47]Reznek, op. cit., p. 77. Support for this idea is also found in Campbell, E. J. M., Scadding, J. G. and Roberts, R. S. (1979) The concept of disease, *Br. Med. J.* **29** (September), pp. 757–762.

[48]Fabrega, H. Jr., in Concepts of disease: logical features and social implications, in Caplan et al., op. cit., pp. pp. 493–522, speaks of a phenomenological framework for describing disease. "The defining characteristics of a disease formulated within this framework would include changes in the *states-of-being* (e.g., feeling, thought, self-definition, impulses, and so on) which are *(a)* seen as discontinuous with everyday affairs, and *(b)* believed to be caused by socioculturally defined agents or circumstances." p. 505. Folk beliefs and symbol systems would structure these beliefs, e.g., disease as punishment. Again, these beliefs will ground a distinction between defining characteristics of disease (it is a punishment) and its indicators (the symptoms), but Fabrega seems to confine this analysis largely to preliterate societies. My claim is that it should apply to our own society also.

[49]Lewis, G. (1993) Some studies of social causes of and cultural response to disease in Mascie-Taylor, C. G. N., ed., *The Anthropology of Disease,* Oxford University Press, Oxford, p. 104. For other interesting examples, *see* Staiano, K. V. (1986) *Interpreting Signs of Illness: A Case Study in Medical Semiotics,* Mouton de Gruyter, Berlin.

[50]Hammer, L. (1990) *Dragon Rises, Red Bird Flies: Psychology, Energy & Chinese Medicine,* Station Hill, New York, p. 389.

[51]Zaner, R. M. (1982) in Chance and morality: the dialysis phenomenon, in *The Humanity of the Ill: Phenomenological Perspectives,* Kestenbaum, V., ed., The University of Tennessee, Knoxville, pp. 38–68, has argued that this undeserved nature of affliction is the basis of the moral imperative to assist those who suffer from disease.

[52]Sontag, S. (1978) *Illness as Metaphor,* Farrar, Straus & Giroux, New York, and (1989) *Aids and its Metaphors,* Farrar, Straus & Giroux, New York.

[53]Sontag, 1978, op. cit., p. 73. Another relevant quote: "The formal theory of illness of a group, together with folk understanding, gives a distinctive ideological cast to what can be termed the group's medical care system": Fabrega, H. Jr., The scientific usefulness of the idea of illness in Caplan et al., op. cit., pp. 131–142, p. 139.

[54]Calnan, op. cit., p. 139, gives the following quote from an interview:
Q: How has your health been over the last few years?
A: ... apart from the fact that I am a disabled person, and apart from my aches due to an accident, it is excellent.

[55]Epictetus said the following: "Ask not what events should happen as you will, but let your will be that events should happen as they do, and you shall have peace." And he follows this aphorism immediately with: "Sickness is a hindrance to the body, but not to the will, unless the will consent. Lameness is a hindrance to the leg, but not to the will. Say this to yourself at each event that happens, for you shall find that although it hinders something else it will not hinder you." Paragraphs 8 and 9 of (1966) The manual of Epictetus, Saunders, J. L., ed., in *Greek and Roman Philosophy after Aristotle,* The Free Press, New York, p. 135.

[56]Caplan et al., op. cit., p. 41.

Introduction

This paper elucidates the concept of disease as found in the practice of alternative or complementary medicine, partly through a comparison with the biomedical conception of disease. Each paradigm, it is argued, conceptualizes disease in overlapping, but divergent, ways. As a result, traditional and alternative medicine should not be interpreted as merely diagnosing and treating the same thing in different ways. Indeed, the unfamiliar and even bizarre modalities associated with alternative medicine only make sense (irrespective of their effectiveness) in light of the more inclusive definition of disease that underwrites the practice of alternative medicine.

The Concept of Disease in Alternative Medicine

Mark B. Woodhouse

My purpose in this essay is to elucidate the concept of disease (and, by implication, the concept of health) in the paradigm of holistic medicine. This paradigm also carries the label of "alternative" or sometimes "complementary" medicine. These are not synonymous terms; an alternative treatment may or may not complement conventional medical practice, and a holistic paradigm of health includes far more than a summary of alternative treatments. However, since these terms are closely associated in popular and professional thinking, I shall treat them more or less synonymously for purposes of this essay. Since the holistic conception of disease is not well understood, I shall undertake its elucidation partly through a comparison of key claims with the more familiar biomedical model.

After some brief historical remarks, I shall address the following topics: the definitions of disease and health vis-a-vis the holistic and biomedical models; diagnostic methods (because they illuminate the concept of disease); treatment methods (because they also illuminate the concept of disease); ethical issues relating to patient/provider responsibilities; and the underlying physics and metaphysics of disease in the holistic model.

Before getting underway, let me clarify several working assumptions and offer a brief historical overview. In the interests of clarity, I shall describe somewhat idealized versions of the holistic and biomedical models as they apply to health and disease. In practice there is growing crossfertilization between these paradigms and, in the interest of patient safety, health providers naturally may be expected to qualify and revise their working assumptions in ways too numerous to describe here.

For purposes of illustration I shall select treatments from alternative medicine for which there is clinical or experimental evidence. Of course, evidence is not the same as proof, but this essay is not the place to debate the merits of particular treatment modalities, the appropriateness of experimental protocols used to evaluate them, or the politics of medicine, which sometimes plays a role in such debates. My purpose is rather to elucidate the conceptual parameters of disease according to holistic medicine.[1]

Many treatment modalities and a few of the diagnostic procedures associated with alternative medicine have existed for hundreds and even thousands of years in various cultures throughout the world. Acupuncture and herbal medicine have been part of Chinese culture since before Confucius. Ayurvedic medicine, which diagnoses disease although subtle differentiations of pulse movements, predated the Aryan invasions in India.

The Aesculapian healing temples of ancient Greece utilized dream therapy and psychic diagnosis, as well as herbal remedies. Ironically, Hippocrates—sometimes described as the "father" of modern medicine—actually believed that a properly functioning spinal column was the key to good health, a claim most often associated today with chiropractic practice.

Many holistic practices today incorporate a theme central to the practice of shamanistic traditions the world over—African, Siberian, Indonesian, Brazilian, and Native American. This is the belief that physical disease reflects, on a deeper level, spiritual imbalances. Paracelsus, a pioneering surgeon, also prescribed alchemical potions for his clients. Mesmer, the developer of hypnosis, also practiced what is today described in holistic circles as "magnetic healing," and Samuel Hahnemann, who 200 years ago discovered homeopathic remedies, was himself a distinguished practitioner of allopathic medicine, which is today foundational to the biomedical model.

Interest in alternative medicine is growing exponentially. According to a study in the *New England Journal of Medicine* (January 28, 1993) Americans spent $14 billion on alternative remedies in 1992. More than 80% was out-of-pocket, and more than 50% was in addition to standard medical procedures. Several health insurance companies, among them Prudential, New York Life, and Metropolitan Life, have undertaken pilot studies to determine if, with selected diseases, such as arteriosclerosis, alternative medicine would be cheaper and/or more effective. Moreover, under a Congressional mandate, the National Institutes of Health have established a special office devoted entirely to the clinical investigation of the effectiveness of alternative strategies. Over 25 major medical schools currently offer courses, clinics, or modest research programs in alternative medicine, and Harvard Medical School recently awarded the first AMA-sanctioned continuing education credits in alternative medicine for physicians.

Despite this emerging interest, however, the average physician or philosopher is not well informed about the diagnostic and therapeutic modalities of holistic medicine, or about its underlying assumptions. Those that are modestly informed often have developed negative attitudes out of proportion to available empirical evidence or client circumstances. Some of these attitudes reflect understandable ethical concerns over client well–being, and some, I am afraid, reflect *a priori* biases against developments that do not fit the prevailing paradigm. What is it, then, about the holistic conceptions of disease and health that fuels the current controversy?

The Relationship of Health to Disease

It is often assumed that holistic and biomedical models share more or less identical conceptions of health and disease and differ only in their treatment approaches. According to this assumption, the alternative practitioner may prescribe, say, acupuncture, a radically altered diet (supplemented with herbal remedies), and psychotherapy for cancer, whereas the traditional physician will prescribe chemotherapy and such dietary (more fiber) or lifestyle (no smoking) changes as may be directly related to the type of cancer involved.

In fact, however, holistic and biomedical models conceptualize health and disease in quite different ways. These conceptions partly overlap in some instances. In others, the differences are contingent and may be bridged by further research. In still others, the differences are so logically incompatible that only a conceptual revolution can bridge the gap. These differences guide and inform the respective treatments of choice. Therefore, rather than differentiating holistic and biomedical models on the basis of their treatment modalities, understanding why the treatments can vary so radically in the first place requires that we understand how the concepts of health and disease vary in these two models.

With some notable exceptions, such as immunization programs, the biomedical model is built around the detection and treatment of disease, and leaves health largely undefined and underemphasized. Although everyone obviously wishes to "be healthy," the concept of health in the biomedical model functions essentially as a default value. That is to say, health is what is "left over" when a person is free of all disease and obviously predisposing factors to disease. Physicians are trained, first and foremost, to diagnose and treat disease and only secondarily to promote health in all but the most rudimentary arenas of prevention, such as smoking and prenatal care.

Most, of course, believe that they are promoting health by helping their clients to overcome disease, practicing preventive medicine, and supporting public health programs. Taken as a summary of successes with particular diseases and symptoms of disease, I cannot imagine anyone disagreeing with this assessment, yet physicians typically do not actively promote their patient's health in general; they apply instead methods for dealing with particular diseases—all the more so in the specialized branches of medicine.

One reason physicians do not promote health as readily as they attack disease is that the biomedical paradigm does not offer a robust conception of health to begin with. When asked, they may attempt to diminish this conceptual vacuum by appealing to test–levels (e.g., liver function) that are arbitrarily based on population norms, irrespective of whether members of that population are truly healthy in anything more than the circular sense that they are not sick. The fact that one's liver function is "within normal parameters" does not necessarily mean that one's liver is healthy.

By contrast, the holistic model attempts to fill this gap between disease and health with as much empirical detail and prescription of optimum levels as possible. It fills in the "preventive" in preventive medicine with more concrete strategies than the biomedical model typically accommodates and offers a more robust description of health.

According to the holistic model adopted by the World Health Organization, disease is the default negative value resulting from a failure to maintain optimum health. In this view: "[Health] is not merely the absence of disease or infirmity . . . [but] the ability of a system (for example, cell, family, society) to respond adaptively to a wide range of environmental challenges (for example, physical, chemical, psychological, and so on)." Disease is a failure of adaptive response ". . . resulting in disruption of the equilibrium of the system."[2] Consequently, the way to treat disease is not so much to attack the symptoms or the immediate underlying pathology as it is to restore the patient to a state of overall health.

Here, then, is an important difference of emphasis to take into account, even if overly simplified at this early stage of elucidation. In the biomedical model, health is left over when you have cured or prevented a patient's disease, whereas in the holistic model disease results from not maintaining optimum health. This divergence at first looks like a merely verbal difference in the same class with asking whether the glass of water is half empty or half full. It looks as if I am stressing, at best, a mere tautological difference worded in different ways; if a person is not sick, she is healthy, and if she is not healthy, she is sick. Stated like this, the difference would be vacuous, if the same definitions of disease were at stake, but they are not.

Different Conceptions of Disease

The biomedical model defines disease generically in terms of negative symptoms plus (where it is known) the directly related physical pathology (genetics, viruses, inability of an organ to maintain homeostasis, breakdown of structural or biochemical pathways, and so on) that causes those symptoms. The concept of a "negative symptom" is open-ended, but typically may involve pain, weakness, discomfort, nausea, or the inability to perform certain standard tasks. To take a simple example, to have the "flu" is to have a certain flu-causing virus in one's bloodstream, plus the symptoms involving high temperature, weakness, perhaps cramps or nausea, and irregular digestion and elimination.

In the biomedical model, the definitional focus of disease is on the symptoms; you can have flu-causing germs in your system, but if you do not have the symptoms of the flu, you do not have the flu. You can have HIV in your bloodstream, but you do not have AIDS until you have deteriorating physical symptoms. Indeed, one can suffer from a certain disease and not know what the causes are, as is the case, for example, with "Desert Storm Syndrome."

Still, the biomedical model cloaks an ambivalence in this regard; it often terms the negative symptoms involved as symptoms of the disease, as if the disease were the cause (and should be so defined) of the symptoms—not roughly equivalent to them. I have, therefore, included both the underlying pathology and the symptoms in the biomedical conception of disease, with the qualification that whether one or the other is stressed may vary from one disease to the next.

The variable emphasis between symptoms and causes of disease in the biomedical model sets the stage for radical conceptual revision in the holistic model, because not only does it not define disease in terms of its symptoms, neither does it think of disease merely in terms of underlying proximate physical causes. Rather, it defines disease as a breakdown, block, or imbalance somewhere else in the systems, subsystems, and metasystems that comprise an otherwise healthy person, including the person's relationship with physical and social environments.

Symptoms are not the disease, but merely evidence that such a breakdown has occurred. Holism seeks to push the causal chain of factors leading to this imbalance into territory typically left uncharted, and sometimes unimagined, in the biomedical model. It does not deny the efficacy of the necessary and/or sufficient physical conditions stressed by the biomedical model, e.g., HIV is necessary but not sufficient for AIDS. Rather, it seeks to identify and treat a wide spectrum of predisposing factors that individually may seem irrelevant to the patient's symptoms, but that collectively can lead to disease.

The biomedical model is often characterized as reductionist, and holism is typically characterized as logically opposed to reductionism. This suggests that the holist might reject the biomedical ("lower end") conception of disease and attempt to replace it with his or her own. This, however, is not what holism prescribes. Rather, it seeks to include all that the biomedical model characterizes as disease, but then expand this concept to include a wide spectrum of predisposing factors that the average medical doctor typically (although not necessarily) has neither the time, the interest, nor the training to explore.

What drives this expansion, from the holistic practitioner's point of view, is the *a priori* assumption that underwrites all versions of systems theory, to wit: Every part of a system is directly or indirectly related to all other parts, and every system is directly or indirectly related to every other system. No system stands alone. It is all a matter of where, for practical reasons, we wish to establish certain provisional boundaries and consider certain parts or certain systems in isolation from others. The recent growth of environmental medicine testifies to this fact.

In general, the biomedical model draws narrower boundaries, whereas the holistic model draws wider boundaries. (This is why the holistically oriented physician has an information form five times larger than the traditional provider's for an initial consultation.) Roughly speaking, what is distinctive about alternative medicine is determined by subtracting the narrower domain from the broader one. However, most holistically oriented physicians caution that this does not establish any kind of exclusive disjunct between the two paradigms; the practice of holistic medicine by its very nature must include many features of the biomedical model. If it did not, it would not be "holistic." Let me now provide some empirical content for this conceptual distinction.

As a simple illustration, consider the case of the flu. We encounter the typical symptoms (headache, upset stomach, elevated temperature, and so on), plus (if it is a known strain) the specific virus that is causing these symptoms. Together, they constitute the disease we informally call the "flu." This territory is familiar to both the traditional and holistic practitioner. However, another factor is directly related to the onset of the flu, namely, the failure of the immune system to prevent this virus from "taking hold" and causing the symptoms we experience.

Why did the immune system fail in this case? It is easy to respond as follows: "Well, it just did. It was probably overworked and confronted something beyond its ability to deal with. Besides, the important thing is to get rid of the flu, and we can create an immunity to future flu of this type by taking a flu shot." Several factors about this response are worth stressing.

In the first place, it focuses on finding a specific remedy for a specific disease, rather than improving the health of the immune system in general. Dramatic short-term results are what we tend to value. Of course, in overlooking the bigger picture, we have taken so many injections for so many different kinds of flu that we are in danger of losing our resistance altogether; viruses keep mutating as our immune systems gradually lose their innate capacity to respond in creative ways. By comparison, unless there is a life-threatening situation, the holist typically suggests not attacking it with quick fixes, but rather, giving the immune system a workout.

Second, all of the interconnecting factors that can negatively affect the general health of the immune system are generally ignored or set aside in the above response. For example: nutritional deficiency, electromagnetic field exposure, toxic environment, geopathic stress, emotional stress, insufficient rest, prior illness, subclinical allergic reactions, and the progressive inability to synthesize essential precursor hormones for optimum immune system response, to name a few.

Physicians are paid to get rid of our flu, or at least to relieve its symptoms. We do not as a rule pay them to investigate the general health of our immune systems. Yet, the general health of our immune systems is precisely what needs to be examined in this age of increasing autoimmune disorders. It is not that the holistic practitioner is unconcerned about relieving symptoms; she usually has a variety of remedies to recommend, depending on the illness. It is just that with the individual who frequently contracts the flu, the symptoms signal deeper issues that affect the immune system itself.

Returning the immune system to a state of optimum efficiency takes time, but doing so will prevent future bouts with the flu just as efficiently, if not more so, as flu shots; it will also pay handsome dividends in warding off other potential diseases; and it will not lead to the self-defeating consequences now faced by public health officials who have taken a piecemeal approach to contagious diseases for so long. Therein is a vitally important difference between holistic and biomedical approaches to health and disease.

According to the holistic model, most people are more or less in a continuing state of moderate ("subclinical") disease, even although they may not be manifesting symptoms of any particular disease. They are in these states of disease to the degree they are not in a state of optimum health. We typically only apply the label of "disease" when we are so far removed from optimum health that specific symptoms are encountered, but if our immune systems are working only moderately well, then we are moderately diseased, even if there are no symptoms of specific disease.

The manifestation of the symptoms is simply the lowest point on this scale at which one is most diseased, i.e., the farthest from optimum health. Once they manifest, the holist insists that the symptoms, plus the inner physical causes, and everything that contributed to the patient's inability to resist disease, are all part of the "disease."

In the biomedical model, one can understand disease without a prior understanding of health, because the latter is parasitic on the former. In the holistic model, as I have portrayed it, one cannot understand disease without a prior understanding of health, because the former is parasitic on the latter. My comparison thus far is modeled in Fig. 1.

By "optimum health" the holist understands a more or less permanent condition in which all relevant systems, all parts of those systems, and all interactions between systems are functioning at their optimum levels. Optimum health also means maximum resistance to disease. Thus, an individual who consumed higher than average amounts of cholesterol, smoked a pack of cigarets a day, and carried HIV virus, all for 20 years, but who also showed no signs of heart or arterial disease, lung or breathing

		Intermediate zone: Neither sick nor especially healthy	
Standard clinical factors are in evidence	Subclinical factors relating to disease, including psychosocial ones, are detectable but often ignored or overlooked.		
Symptoms (A)	Pathology (B)	especially healthy (C)	Optimum Health (D)
Patient meets criteria for a specific label of sickness and therefore is ill.		However, the patient has no complaints, feels OK, and there- fore is assumed to be healthy.	

Biomedical model
Disease = A + B
Health = C
D is acknowledged as
a logical category,
but left undeveloped.

Holistic model
Disease = A + B + C
Health = D
C commands much
of the practitioner's
skills.

Fig. 1. The closer to (A), the more individually necessary and/or sufficient etiology becomes; the closer to (D), the less so.

disorders, or AIDS symptoms, or any other signs of disease, may be said to be, if not optimally healthy, certainly healthier than the individual who succumbs to one or more of these diseases in the same time period.

Conventional medicine asks: "Why do people contract various diseases, and what can we do to reduce their occurrence?" Although certainly concerned about these questions, the holistic physician additionally asks: "Why do some people remain healthy throughout their lives, despite being exposed to the same factors as those who become sick?" This global approach is the key to prevention, rather than a piecemeal approach to each disease; hence the use of "holism" to characterize this approach.

I cannot begin to describe all that the holist seeks to incorporate into this conception, a lack of which begins the long journey toward the tangible symptoms the traditional physician labels as a specific disease. What I can do, however, is describe three areas that illustrate the kind of thinking that informs this expanded conception of health and therefore helps to distinguish it from the biomedical model.

Colon activity: From studies with cadavers, it appears that the average 45-year old male carries 5+ lbs. of impacted fecal material "glued" to the sides of his intestines. This can include, for example, small portions of undigested pizza consumed 10 years ago. The average person also defecates only one time per day. Neither situation is "healthy" from the holistic point of view—and not merely because, as a result of poor dietary habits, the colon itself may become diseased.

Poor-to-average elimination allows toxic substances to build up in many other systems of the body, because they cause the liver to be overworked. The buildup takes place over decades and contributes to premature aging. According to the biomedical model, the internist should check the patient's stool once a year for blood. An extended check for intestinal polyps may also be indicated. If the results are negative on both counts, the patient is deemed "healthy" in this regard.

From the holistic point of view, however, this is only the proverbial tip of the iceberg. The patient may well be on his way to an early death, even though there are no tangible symptoms, and those that do surface, such as cracked toenails and heavy callous buildup on the feet, are typically not connected to a weakened liver. On this count, the patient is definitely not healthy and therefore, is what we might describe as "subclinically diseased."[3]

Nutritional biochemistry: According to the biomedical model, if we eat moderately, reduce fat and cholesterol intake, include five servings of fruits and vegetables, and take perhaps one vitamin/mineral supplement per day, then barring any specific deficiencies (such as calcium depletion leading to osteoporosis), we are nutritionally healthy. From the holistic perspective, this is a useful prescription for someone who lives on meat and potatoes, but barely scratches the surface of genuine nutritional health.

The key here is "bioavailability." Not only should often missing nutrients be identified, but all nutrients also must be consumed in adequate amounts, in the correct proportion relative to each other, and at the correct times. Moreover, the client's enzymatic and hormonal functions must be taken into account in determining how much of the nutrient is actually being absorbed into the system, first of all, and then how much is actually utilized by the various subsystems of the body. Cereal box claims that, with a bowl of their product, we are receiving 100% of various vitamins and minerals that we need on a daily basis are grossly misleading.

It has been observed that we are a nation of starving people with generally full stomachs. The average person does not assimilate anything near the amount of nutrients that are initially consumed; they are merely passed on or, in some cases, built up to unhealthy levels, because bioavailability varies substantially from person to person and nutrient by nutrient—and sometimes even by brand. There are quite literally hundreds of major interactions to take into account, and thousands of minor ones. To take a very simple example, the average person believes that having a glass of orange juice is a good way to obtain vitamin C.

From the holistic perspective, this is, of course, better than nothing. However, far more vitamin C would be made available to the system if we ate an orange instead, because the pulp in the orange contains other enzymes and micronutrients that support its chemical breakdown and assimilation. Because the average person is so far from optimum bioavailability, that person is nutritionally diseased according to the holistic perspective even though, again, there may be no symptoms of disease from the biomedical perspective.[4]

Emotional/spiritual well–being: Holistic and biomedical models differ in two important respects in this category. First of all, according to the biomedical model, if one is not depressed, guilt-ridden, obsessive, paranoid, and so on, then it is assumed one is psychologically "OK." However, as everyone knows, and the holistic model stresses, there is a large difference between merely being not depressed, on the one hand, and being genuinely happy, on the other.

The holistic model holds out ideals for daily living that are well beyond what most of us typically experience, among them joy, peace of mind, high energy, empowerment, ability to process high stress efficiently, and the capacity to express emotions freely. Compared to these ideals, most persons are emotionally unhealthy—and therefore, emotionally diseased—according to the holistic model.

This brings us to a second important difference. Until very recently, a person's emotional health and well-being was considered essentially irrelevant to his or her physical health and well-being, according to the biomedical paradigm. The emergence of psychoneuroimmunology (facilitated by the discovery of how neuropeptides work), however, has implicated guilt, hopelessness, stress, anger, and lack of forgiveness in a variety of diseases, especially cancer.[5]

Holistic practitioners consider this a welcome development, but carry matters much further. According to their model, all disease has underlying emotional components that predispose one toward specific diseases. Whether disease results, of course, depends in part on the degree of emotional unhealth and many other factors of a physical sort, such as environmental toxins or genetic predispositions.

This grandiose claim is a direct consequence of the holistic conception of human nature; all systems are interconnected. This view has only modest empirical support at the current time, but in fact lends itself to empirical test in the future on a disease-by-disease basis. In the meantime, holists conceptualize matters thusly: Lack of optimum emotional health, over the long term, contributes to physical disease, and to the extent that most of us are not nearly as emotionally healthy as we could be, we are in this arena "subclinically diseased."

The three areas I have selected to illustrate the differences between holistic and biomedical conceptions of health and disease are but a few of the hundreds available. Sufficiently generalized, these kinds of considerations in the holistic paradigm lead to a rather startling conclusion. According to the biomedical model, which interprets the lack of specific disease symptoms and underlying (physical) pathology as grounds for health, most of us are healthy most of the time. According to the holistic model, most of us are unhealthy, and therefore, diseased, most of the time.

The Diagnosis and Treatment of Disease

Contrasting approaches to the diagnosis of disease logically follow from the diverging conceptions of health and disease just described. The biomedical model offers a diagnosis based on manifest symptoms plus the underlying proximate physical pathology. Let me stress here that the holistic model is entirely consistent with, and encompassing of, such diagnoses; there is no reasonable way to deny what is a matter of empirical fact. If one has all the symptoms of cancer, including a positive biopsy, then one has cancer.

These contrasting approaches to diagnosis, however, are asymmetric. Holism includes everything that the biomedical model would have to say regarding cancer, but then goes in search of additional components that directly and indirectly contribute to the onset of cancer. The biomedical model, by comparison, does not include the holistic one in this regard.

Why, one might reasonably ask, is there a need to go beyond the diagnosis offered by the biomedical model? After all, cancer is cancer. The answer is pivotal to understanding the current debate around holistic medicine. Briefly, it is this. How one diagnoses a disease is directly related to how one conceives the range of treatment modalities; to treat the disease is to treat what one has defined as the disease.

If cancer is merely its overt symptoms and pathology, and nothing more, then that is what one must treat. On the other hand, if cancer is its symptoms and its pathology, plus a spectrum of other clinical and subclinical predisposing factors, then one's treatment options are magnified accordingly.

There is another reason why this distinction between restricted and expanded diagnoses of cancer is critical. If one acts only on the restricted definition of the disease, and fails to address the broader range of contributing factors, then apparent success in treating the symptoms leading to a remission will leave open the probability of recurrence several years later. In this event, the original treatment will be described as only "partly successful"— meaning that it failed to do what everyone hoped it would do. One of the reasons it failed, from a holistic perspective, is that the diagnosis of cancer was so narrowly proscribed in the first place. The surgeon who "gets it all" is understandably pleased. The alternative practitioner who sees the patient returning to a life of despair is not as optimistic about remission.

It should be emphasized that the holistic practitioner is not drawing attention merely to the obvious predisposing factors, such as smoking in the case of lung cancer or fat and cholesterol in the case of heart disease. Dozens of other factors already alluded to both may predispose one toward cancer and simultaneously detract from optimum health. Many may seem to have nothing to do with cancer per se, but instead weaken the immune system's ability to resist cancer, as well as other diseases. The holist is more concerned with the multiple factors that contribute to overall health, but that taken individually are often neither necessary nor sufficient for the occurrence of specific diseases.

Before continuing, let me address an objection that earlier may have attracted the reader's attention. Briefly, it is that my depiction of matters has conflated questions of meaning with issues regarding causality. The question of what cancer (or any disease) is, it will be observed, should be distinguished in principle from the question of what causes cancer (or any disease). My response to this objection relies on the fact that the boundaries between synthetic and analytic statements are open to change.

Let me point out, first of all, that even within the parameters of the biomedical model, the boundaries between meaning (the definition of a certain disease) and causality are often blurred. A malignant tumor, for example, may be both the cause of other symptoms relating to brain cancer, e.g., dizziness, and the basis for the primary diagnosis (and hence definition of) the cancer. The lines between a patient's experienced symptoms, the underlying pathology, and the factors leading to that pathology are difficult to draw, because in reality all three segments are part of a continuum that both ordinary and medical language allow us to "carve up" at somewhat different points.

Both symptoms and underlying proximate causes, such as a virus, inform the definition of specific diseases in the biomedical model, with no sharp distinctions between them. Viewed in this light, the holistic practitioner merely wants to extend the definition of specific diseases to nonproximate causes so far as to become continuous with a definition of disease in general.

Where we draw the line between the factors that inform the definition of particular diseases, particular classes of disease, and disease in general, on the one hand, and the factors that we take to be in some way causally implicated in any of the preceding, on the other, is not an objective process. It is, rather, a conventional one that can be affected as much by politics and economics as by science.

Since the holist is stipulating a greatly expanded conception of disease, he does not have ordinary language (which to some extent is already allied with the biomedical model), on his side. So long as he makes it clear—as I am attempting to do in this essay—that he is proposing, for plausible reasons, an expanded conception, he is not subject to the objection of having confused questions of meaning with those of causality.

As noted earlier, there is a powerful tendency to think of biomedical and holistic treatment modalities as essentially different treatments for the same thing. In some instances this is true. Yeast infections, for example, can be effectively treated by pharmaceutical preparations or by grapefruit seed extract. In most instances, however, this assumption is either false or at best only partly true, depending on how broadly or narrowly one describes "the same thing." Under the narrow conception of cancer (symptoms plus pathology), for example, this assumption is false. Under the broad conception, it is partly true. Here is why.

The treatment of preference for cancer under the biomedical model is some combination of surgery, radiation and/or chemotherapy, plus eradicating any lifestyle factors that directly contribute to the cancer. A support group may be recommended, but mainly to help one deal with the emotional challenge brought on by the disease (e.g., fear of death). The average oncologist does not believe that presence or absence of love in a patient's life, for example, has anything to do either with the causes or the cures for cancer. Thus, he focuses on treating the specific manifestation of cancer, usually tumorous growth, with drugs and surgery.

To use the battle metaphor, the holistically oriented physician, although not necessarily rejecting the aforementioned treatments (depending on the stage of the disease), "goes after" a whole different segment of the health–to–disease spectrum. She may prescribe acupuncture (which is ostensibly used to "unblock" the flow of 'ch'i' energy along certain meridians associated with healthy lungs and optimum immune system response); nutritional analysis and support (in which, for example, it may be discovered that certain critical phytochemicals are missing that hormonally support the optimum function of the immune system); and psychotherapy (not merely to help the patient "deal" with the illness, but to undertake certain core emotional release work around, say, the lack of forgiveness and guilt associated with childhood trauma, which has subsequently restricted the ability to give and receive love).

None of these treatment modalities directly fight cancer *per se*. All indirectly fight it by supporting other systems in the background, the optimum function of which will support the fight with or without the use of traditional surgery and chemotherapy. Surgery may well be necessary as an immediate counter measure, but in the overall long–term view of matters, the holist proceeds on the assumption that it is best to restore all of the patient's own natural systems for fighting disease to optimum efficiency.

This is why, from a traditional perspective, the holist's treatment modalities appear to have nothing to do with fighting cancer *per se*. They are less locally aimed at the tangible manifestations of cancer, but instead are nonlocally "spread out" to other systems of both higher and lower order complexity throughout the patient's body and brain/mind. The more nonlocally spread out they are, the less individually necessary and/or sufficient they will be for particular diseases, although collectively they predispose one toward diseases in general. Thus understood, biomedical and alternative treatment methods are not merely different procedures aimed at the same target. They are different procedures aimed at different targets.

Another way to frame the different approaches of holistic and biomedical models to the treatment of disease is as follows. Under the biomedical model, treatment procedures do not parallel the causes of the disease except in the relatively few cases for which lifestyle changes are recommended, e.g., obtaining more exercise. No matter what caused the cancer, the principal treatment methods—surgery, chemotherapy, radiation—are unrelated to those causes. They are aimed instead at what those causes, whether known or unknown, actually produced.

Under the umbrella of alternative medicine, by contrast, treatment modalities generally do parallel the various causes that have indirectly conspired to allow the cancer to happen. It is not that holistic practitioners are indifferent to the client's taking in known carcinogens; their recommendations would be identical with those of traditional medicine in this regard—avoid them like the plague!

Their principal concern, however, is with the background support systems that have become so weakened they have allowed carcinogens to become causally efficacious.

Although it is a bit of an oversimplification, let us assume that (per typical holistic thinking) a patient's diminished capacity to effectively use antioxidants, repressed anger, blocked flow of 'ch'i' energy, and sluggish T–cell production are the main culprits in her lung cancer. She also smokes moderately. Smoking is certainly implicated in lung cancer. Yet, as everyone knows, some people smoke and never develop lung cancer. The difference is in the background support systems. Had the aforementioned factors not been in evidence, but instead signaled a high degree of resistance to all disease, she would in all probability not have developed cancer—whether or not she smoked moderately!

Thus, the holistic practitioner's treatment regimen will parallel many of the factors that allowed the patient to become vulnerable in the first place. The nutritionist will go to work uncovering why the patient's own antioxidants (plus others in her multivitamin/mineral supplement) had such a low degree of bioavailability; the acupuncturist will uncover the meridians at which ch'i energy is blocked; and the psychotherapist will uncover the repressed anger that is, not altogether metaphorically, "eating her alive." Each of these procedures is directed at the enabling causes of the cancer and in that sense parallels them. In holism, treatment parallels etiology; in conventional medicine, it does not.

Patient/Provider Responsibilities

The holistic conceptions of health and disease have the potential to radically alter the way we parcel out doctor/patient responsibility for health care. Under the biomedical model, disease tends to be so narrowly defined in terms of its most critical manifestation that only a physician with the requisite years of training is licensed to undertake the appropriate diagnosis and treatment.

Despite advances in informed consent and in obtaining second opinions, however, this still tends to leave the patient somewhat at the mercy of the system. It is inherently hierarchical in that the doctor will virtually always know which procedures it is best to follow, of those sanctioned under the biomedical model. This is all the more the case in highly specialized arenas of medicine. The patient has come at a time of vulnerability, namely, the onset of illness or disease, and is hoping or expecting that the doctor will be able to "fix" things. The doctor wields a considerable amount of power in such situations—and rightfully so. Training and experience give him or her that right.

However, under the holistic model, in which disease is more broadly conceptualized to include its distantly related enabling causal roots, there is much more that the patient can do for him- or herself. Any person of average intellectual ability can read about these contributing factors and undertake a program of optimizing his or her health, and public schools can begin to educate students about them. The same person can make reasonably informed decisions about a whole range of treatment modalities and self–prescribe accordingly—ideally before disease strikes!

In this view, the holistically oriented physician is more of a partner in the creation of health, rather than a court of last resort after disease has set in. Most readers of this article, for example, are capable of improving the strength of their immune systems without the assistance of a physician, although a holistically trained doctor will be able to make a variety of useful recommendations in this regard.

A holistic paradigm of health and disease shifts a greater responsibility not only for health maintenance, but also for treatment of disease, from the provider to the patient. It gets the patient more actively involved in managing his or her health and disease, whether in undertaking visualization exercises or researching Chinese herbal remedies. It thereby reduces the imbalance of an otherwise more hierarchical relationship between physician and patient so common in the practice of mainstream medicine. There are obvious pitfalls in making this shift, which I cannot examine here. What I hope to have shown is why, under the holistic model, it is a natural one to make.

An interesting ethical and, to some extent, sociopolitical issue has surfaced in connection with this shift in perceived responsibility for health. Briefly, it is this. If a person is educationally and financially able to undertake certain practices that would both substantially increase resistance to disease (through the creation of optimum health) and treat it more effectively after the fact (by attending to the diminished capacity of background support systems), then, if he fails to implement this ability, to what extent should we say that he is responsible for, indeed, has literally created, his own disease?

This issue has become significant for psychoneuroimmunology, in which unhealthy emotions and attitudes are causally implicated in physical disease. In particular, some critics of holistic medicine associate it with a "blame the victim" mentality that, in their view, is completely unjustified. We should be extending compassion for a cancer patient's pain and suffering, together with offering the best treatments available. We should not attempt to trace the causal origins of the cancer partly to some alleged character defect in the patient, such as her inability to express anger, thereby blaming her for creating her own disease. From the perspective of one whose life is devoted to attempting to cure disease, this is a reductio ad absurdum of the holist's conception of disease. Indeed, many holists accuse each other of this transgression.[6]

I think this is a bogus issue for several reasons. In the first place, it may be brought on by the questionable assumption that we ought to hold persons responsible only for their overt behavior, such as smoking too much or failing to exercise, but not for the attitudes or desires that prompt such behavior, at least insofar as health and disease are concerned.

Holists tend to reject this assumption. If poor diet, excessive smoking, and inability to express anger are to varying degrees causally implicated in lung cancer, and if a person is in the position to change each of these factors but fails to do so, then to that extent he or she is responsible for the cancer.

It is at this point that the logic employed by the critic of holism becomes questionable, because he assumes that to be responsible for X means just to be morally responsible and then defines "moral responsibility" as a condition in which the agent deserves praise or blame for a certain action. "Blaming the victim" is the natural outcome of this line of thinking.

The mistake consists in assuming that moral responsibility is the only kind of responsibility possible, at least in contexts of health and disease, when it clearly is not. We are contextually responsible for all kinds of things that have nothing to do with morality, like choosing to take a short cut to work or purchasing a cheaper brand of vitamins.

To be contextually responsible means just that we were neither constrained nor compelled to bring about the consequences that we in fact did. Sometimes this involves moral issues relating, say, to maximizing the common good or to violating another's rights, and sometimes it does not. There is no automatic entailment relation from "Jones is responsible for X" to "Jones deserves blame for X." The charge of "blaming the victim" sometimes leveled at the holistic conception of disease is, therefore, not justified.

Finally, there is nothing about the holistic conception of disease that precludes either feeling or expressing compassion toward the cancer patient. In fact, given the emphasis on personal growth, overcoming negative emotions, and enhancing the capacity for expressing unconditional love that is typical of the holistic conception of emotional health, a cancer patient is as likely, if not more so, to receive a compassionate reaction from the holistic practitioner as from the traditional health care provider. She certainly will not receive blame.

The Physics and Metaphysics of Disease

Holists conceptualize the ontology of disease at a more fundamental level than prescribed by the standard medical materialism. The result is compatible with medical materialism, at least up to a point, but then moves beyond standard materialism in ways that may or may not turn out to be justified. In fact, holists ironically may be interpreted as engaging in an "end-run" around materialists in uncovering the basic common denominator of all disease.

Briefly, the picture is this. Holists take seriously the revolution in physics, whereby matter has come to be understood as uniquely configured patterns of dynamic energy. Physical stuff is quite literally densely compressed energy that differs in degree, but not in principle, from the four forces of nature commonly recognized by physics (gravitational, electromagnetic, and strong and weak nuclear). As David Bohm observes, "matter is simply gravitationally trapped light."[7]

Unlike Arthur Eddington, however, who affirmed that only the "table of modern physics" (colorless, intangible, mostly empty space), rather than the "table of common sense" (solid, visible, enduring), was real, holists do not deny the reality of matter. Holists are antireductionist, not in attempts to affirm the ontological priority of wholes over parts or of consciousness over brain-states, but rather, in their recognition that each level in an ontological hierarchy of complexity and substantiality—from quarks to consciousness—carries its own "reality" and is irreducible to levels below it.

Having said this, however, they want to shift attention away from the behavioral and biochemical correlates of disease toward the underlying energetic correlates.[8] The factors that allow us to differentiate these energetic correlates are field strength, frequency, amplitude, wave form, interference patterns, and phase coherence.

These factors set the stage for the holist's grand working assumption regarding disease: Every organ, physical state, or physical system of a person has a distinctive "vibrational signature" that fluctuates within established parameters, depending on whether the system or organ is healthy or diseased. Determining the appropriate signature can play a role in diagnosis, sometimes before symptoms pierce the veil of clinical discernment. Returning that signature to its optimum (healthy) parameter is the basis of treatment. At this level, holistic medicine is described in some circles as "energy medicine."

Theoretically, the claim that, say, sclerosis of the liver has a distinctive vibrational signature that sets it apart both from a healthy liver and other organs in general, is not implausible. Every element on the Periodic Chart, for example, has its own distinctive wave-frequency configuration, as determined by electron spectroscopy. Magnetic resonance imaging basically coaxes, say, a brain tumor to the point of visual representation through a complex process of pulsed waves that create interference patterns. Kidney stones are pulverized with sonic waves, bearing in mind that if the frequency/amplitude settings are incorrect it is the kidney itself that might be pulverized. Jaundiced premature infants are treated with ultraviolet light.

Why not, then, develop a comprehensive science of energy medicine, for the diagnosis and treatment of disease, to which both the holistically and traditionally trained health care provider can contribute? There is no doubt that both perspectives can overlap to some degree; indeed, they already do, but medical holists also move in directions that the biomedical model either has yet to acknowledge (since the methods, which may or may not be effective, have not been sufficiently researched), or in some cases excludes on the basis of a fundamental incompatibility with the currently known laws of physics. Following are three examples to illustrate the range of this divergence.

First, if we can heal jaundice with ultraviolet light, then why not emphysema? Such is the approach of Sherry Edwards, whose research has led her to conclude that virtually every disease directly and indirectly reflects distinctive (sonic) deficiencies that can be compensated for with relatively simple and inexpensive technology. Edwards claims considerable clinical success with this approach and works cooperatively with physicians to document the results.[9]

This is an example of energy medicine that is not ruled out by physics in general or by the biomedical paradigm in particular and awaits further research. One of the principal challenges in this work is that the detection and compensation of frequencies must be exact in order to be effective.

Second, Western medicine knows that acupuncture works, at least in selected arenas. It does not understand, however, why it works. The reason it does not understand is that, from the standpoint of Chinese medicine, the determining mechanism in health and disease—the flow of ch'i energy through hundreds of meridians in the body (which are unblocked by acupuncture treatments)—is essentially a fifth force in nature; it is not reducible to any of the recognized four forces.

It does appear possible, however, to transduce ch'i energy into a form of energy detectable by sophisticated electromagnetic measuring devices just as, for example, audible sounds are transduced into electrical pulses on telephone lines. Some Japanese hospitals routinely use such devices.[10] Physics, however, rejects the existence of ch'i as a fifth force and holds that whatever is measured by an electromagnetic detection system must be itself of an electromagnetic nature. Acupuncturists and health physicists can collaborate on the diagnostic and therapeutic applications of this electromagnetic interface, but the metaphysics of ch'i remains fundamentally different from the physics of the electromagnetic spectrum.

Third, at the farther end of energy medicine is psychotronics. Here, not only is the energy–information exchange totally off the physical map, but the machines used to diagnose and treat unhealthy conditions are themselves physical anomalies. They are filled with circuit boards and other paraphenalia. They have dials for making fine discriminations of frequency, and the manuals that accompany them often have more than a thousand specific frequency codes for everything from lupus to termites. Yet according to the FDA and the Patent Office, they are inherently incapable of sending or receiving any kind of energy to or from humans, animals, or crops.

Why bother, then, to include psychotronics in the holist's handbag of alternative tools? The answer is because there is circumstantial evidence to suggest that in fact it (or something closely related to it) actually works:

1. Psychotronics is almost a routine adjunct to health care in Russia and former Eastern bloc countries;
2. Otherwise mainstream scientists hold international conferences on the subject;
3. The US military continues to research its potential;
4. There is substantial anecdotal evidence from patients who have gone partly underground (or to Europe) for treatments;
5. A small number of farmers, whose living depends on effective crop maintenance, successfully use psychotronic machines to diagnose and treat everything from corn borers to pH imbalances without the use of pesticides or other chemicals; and
6. The FDA has made it illegal, even for board-certified physicians, to treat patients with psychotronics, which is a strange stance to adopt if such use could never be more effective than the placebo effect of playing with a Tonka truck.[11]

My purpose is not to assess the effectiveness of these or any other methods that have found their way into the energy–medicine category of the holistic paradigm. It has been, rather, to shed additional light on the ways holists conceptualize disease such that their often exotic diagnostic and treatment modalities—which in light of further research may or may not turn out to be as effective as their defenders claim—are not seen merely as exer-

cises in irrationality. If diseases have distinctive vibrational sig-
natures, it makes sense to develop technologies and treatments
that reflect these partly physical and partly (by current standards)
metaphysical articles of faith.

Notes

[1]Readers unfamiliar with the holistic model may find the following
      books useful, all written by MDs: Shealy, N. (1993) *The Creation
      of Health,* Stillpoint, Walpole, NH; Dossey, L. (1994) *Healing
      Words,* Harper Row, San Francisco, CA; and (1986) *Beyond Illness,*
      Shambhala, Boston, MA; Chopra, D. (1987) *Creating Health,*
      Houghton Mifflin, Boston, MA; and (1993) *Ageless Body, Timeless
      Mind,* Houghton Mifflin, New York; Siegal, B. (1986) *Love,
      Medicine and Miracles,* Harper and Row, New York. In addition,
      two new journals, *Alternative Therapies in Health and Med-
      icine* and *The Journal of Alternative and Complementary
      Medicine,* carry generally high quality peer-reviewed studies of
      alternative treatments and extensive discussions of the
      philosophical foundations of various aspects of holistic medicine.
[2]Cited in Caplin, A. L., ed. (1981) *Concepts of Health and Disease,*
      Addison-Wesley, Reading, MA, p. 36.
[3]For discussion and research relating to colonic health, *see* Jensen, B.
      (1981) *Tissue Cleansing Through Bowel Management,* Health
      Systems Press, Escondido, CA.
[4]The literature relating to nutritional health is voluminous. For up-to-
      date research, the *Journal of Clinical Nutrition* carries articles on
      topics, including bioavailability, of significance for both holistic
      and biomedical paradigms.
[5]For a discussion of the roles of imagery, attitudes, and psychoneuro-
      immunology in healing, *see* Locke, S. (1986) *The Healer Within:
      The New Medicine of Mind and Body,* E. P. Dutton, New York;
      and Rossi, E. (1986) *The Psychobiology of Mind-Body Heal-
      ing,* W. W. Norton, New York. For a list of proposed connections
      between emotional health and physical well-being, *see* Hay, L.
      (1984) *You Can Heal Your Life,* Hay House, Santa Monica, CA.
[6]Cf., for example, Dossey's *Healing Words,* Part I.

[7]Bohm, D. (1983) Of matter and meaning, *ReVision* (Spring), p. 34.

[8]This agenda is extensively developed by physician Gerber, R. (1989) in *Vibrational Medicine,* Bear and Company, Santa Fe, NM. across a spectrum of alternative treatments. Becker, R. (1985) *The Body Electric,* William Morrow, New York. is a detailed exploration of the electromagnetic foundations of biological processes. Research into the energetic correlates of health disease, especially relating to spiritual or "energy healing," as it is sometimes called, is reported in the peer-reviewed journal *Subtle Energies.* Included, for example, are studies of healer's attempts to influence bacteria growth at a distance, alter the surface tension of water, and increase the rate of wound healing. Studies are both single- and double-blind. Pioneering work on weak electromagnetic fields is reported by Yale biologist Burr, H. (1973) in *Fields of Life,* Ballantine, New York.

[9]Research reports may be obtained from Edward's research foundation, P.O. Box 706, Athens, OH 45701.

[10]Japanese physician Motoyama, H. (1988) has developed a machine for measuring meridian energy differentials, as described in *Science and the Evolution of Consciousness,* Autumn Press, Brookline, MA. Several types of similar machines are manufactured in Germany and moderately used by alternative practitioners in this country.

[11]Although the title is somewhat misleading and the research dated, Bird, C. and Tompkins, P. (1973) *The Secret Life of Plants,* Avon Books, New York, especially Sections II and III, is still one of the best elementary introductions to psychotronics and carries extensive references to technical journals reporting on work in this field.

Index